A Pearl in Peril

A Pearl in Peril

Heritage and Diplomacy in Turkey

CHRISTINA LUKE

OXFORD
UNIVERSITY PRESS

OXFORD
UNIVERSITY PRESS

Oxford University Press is a department of the University of Oxford. It furthers the University's objective of excellence in research, scholarship, and education by publishing worldwide. Oxford is a registered trade mark of Oxford University Press in the UK and certain other countries.

Published in the United States of America by Oxford University Press
198 Madison Avenue, New York, NY 10016, United States of America.

CIP data is on file at the Library of Congress
ISBN 978–0–19–049887–0

9 8 7 6 5 4 3 2 1

Printed by Sheridan Books, Inc., United States of America

To the boys

CONTENTS

FIGURES

TABLES

FOREWORD

In the summer of 2014 two young women from a small village in western Turkey visited the US Embassy in Ankara. Both were attending universities in Izmir. They had been invited by friends to visit the United States. The friends were archaeologists who had worked for almost a decade in the region where the women had grown up. Their travel plans fell apart when the US Embassy denied their visas on grounds that they could not show sufficient proof that they intended to return to Turkey. They were presumed guilty of flight before they had even been given permission to leave the country.

After consulting with Turkish lobby groups in the US, their second application was more robust. It included land deeds, statements from bank accounts, high school and college transcripts, and letters from the archaeologists (who held positions at US institutions). Their visas were denied a second time. Others familiar with similar situations told the young women that if they applied again and were denied a third time, they would be blacklisted for the next decade. They did not apply again.

During an unrelated meeting at the American Research Institute in Turkey (ARIT) in Ankara, I asked an American diplomat from the US Embassy for guidance about this case. She informed me that this was "unofficial-official" US policy: young, foreign women were a flight risk, and could too easily become a burden on American resources. No public data were available to demonstrate that this was true. In less than the three minutes it took for each woman to present her case to the officer, the US Consular section in Ankara undercut decades of US–Turkish relations and diplomatic practice in the middle Gediz Valley. There was a history here going back to 1947. The Americans had built the roads in the region, and the farmers claimed to owe their sovereignty to the US Marshall Plan, to Marshall himself, a point I made to the American diplomat.

The diplomat dismissed my explanation, "Oh, you are mistaken; the Marshall Plan was never in Turkey." Perhaps feeling less certain, the diplomat reluctantly asked me for a quick lesson in US historical relations in Turkey. This included a small, but significant and targeted, amount of support from the US Marshall Plan. The families of these two young women were perfect examples of what it had set out to accomplish: to make rural farmers, and formerly those who practiced transhumance, prosperous entrepreneurs who would cultivate a set of social practices and opportunities for their children.

This book is not about visas, nor is it about Turkish lobbying in the hallways of US Congress to combat "unofficial-official" policies. Those topics are for a different book. This book is about the Gediz Valley and the landscapes where these young women grew up and went to college, and the deeply rooted perceptions and judgments held by foreign countries and their diplomats about class, identity, assistance packages, and heritage. The US Marshall Plan is only one small part of how the US and other foreign interest groups—public and private—have strategically positioned themselves such that their policies have had—and continue to have—a vast impact on those who live in this region. But Turkish citizens in their early twenties, while they are likely familiar with the significance of the social engineering practices of the Republic of Turkey, are often unaware of just how much their lives have been influenced by policies dictated not only by Turkey but also by foreign entities. These policies have affected access to education and have reshaped cities and rural landscapes.

The US policies and programs that preceded the Marshall Plan, such as the 1933 Tennessee Valley Authority (TVA), and the ideological struggles of the interwar period between Soviet, US, and European visions for Turkey have not been fully understood. The study of how these forces positioned heritage in Turkey is also in its infancy. A lack of coordination among the agricultural and cultural sectors, too, has resulted in growing tensions regarding public and private ownership and, in turn, access to social mobility.

This book opens up these topics for consideration in the context of the Gediz Valley in western Turkey. My decision to unpack these narratives began in May 2011, when I had a fellowship from the National Endowment of the Humanities (NEH) and ARIT to spend a year in Izmir to research heritage management. I was exploring the archaeological site of Bin Tepe and the relationship between farming in rural landscapes and management of archaeological sites subject to the tractor's plow. Looking for information about tractors in Turkey, I came across Richard Robinson's 1952 article, "Tractors in the Village: A Study from Turkey." It focused on US intervention in Adana, but much was familiar to me. Years of conversations with local communities in the Gediz suddenly fell into place.

I realized that many of the roads I drove, the water I drank, and the food I ate was an outcome of US policies. Such development programs were also the primary reasons why heritage had become celebrated, forgotten, even hated. My conviction to tell these stories in a book was further strengthened the day these two young women were denied their visas. It was through my journey with them, their families, and my own family that I realized the degree to which "policies with a purpose" and "diplomatic deliverables" from heritage to agriculture had created an impossible web of entangled misunderstandings and years of mistrust that showed no signs of becoming clearer.

In the face of the rapid development that is engulfing Turkey, driven by forces of globalization from Canada to China, economic policies and legislation promoting highways, trains, mines, and dams have become very fluid. The same can be said for how Turkey has responded to their presentation of heritage on the international stage, notably through UNESCO's World Heritage Committee and its lists. Yet, internally, survey, excavation, and restoration practices in Turkey have become increasingly contested and convoluted. This book begins to uncover the role of the United States in heritage tensions from the late nineteenth century to the present day.

ACKNOWLEDGMENTS

This book is a result of interdisciplinary work that has extended over the last decade. I've pulled together ethnographic and archaeological survey data from work in the middle Gediz Valley, which I co-directed with Christopher H. Roosevelt under the Central Lydia Archaeological Survey (CLAS). CLAS was diachronic in its approach to survey, from Paleolithic to present. In addition to standard archaeological survey, it also explored how recent histories over the last century or so have shaped people's perceptions and reactions to heritage. We worked with those engaged in farming, fishing, and development, from highways to marble quarries. In seeking to understand better archaeological zoning, team members consulted archives. At first these included the mapping and municipal offices in Akhisar, Gölmarmara, Adala, Alaşehir, Salihli, Manisa, and Izmir. Discussions with the State Water Works at the Demirköprü dam, Salihli, and Izmir branches were also extremely enlightening, as were meetings with local tourism offices, directorates of agriculture and livestock in Manisa, and the head mining office in Izmir.

It was through this fieldwork and associated research that I realized the potential for an article. My editor at Oxford, Sarah Humphreville first inspired me to consider how it might come about into a book. Our first conversation was about publishing data from CLAS in a typical survey volume. She asked what else I was working on. I described the initial seeds of this project, and she noticed a dramatic shift in my voice. It was her encouragement that pushed me toward the initial outline and ultimately the final manuscript.

My continued research revealed that much of the major development in the region had been done during the 1930s, and especially the late 1940s to

1960s, often with US funding. This led me to the Turkish National Archives in Ankara and the US National Archives in Washington, D.C., and then to the US National Archives in Atlanta where the records of the Tennessee Valley Authority (TVA) are kept. Expert archivist Maureen Hill was very kind to sort through files. TVA's linkages to interwar tensions led me to the archives of the Le Corbusier Foundation in Switzerland, where archivist Isabelle Godineau was very gracious in assisting me. I also consulted the presidential archives of John F. Kennedy in Boston, the Smithsonian's archives in Washington, D.C., and with the archivist for UNESCO, Adele Torrance and the time of Lynn Meskell, I was able to review key files from the interwar period. From there, I visited the Ahmet Piriştina City Archives and Museum, Izmir, where Director Ayşe Üngör and her staff, especially Onur Eryeşil, were extremely gracious with their time. The networks described in diplomatic cables, personal correspondences, and policy documents housed in these various archives had uncanny similarities to stories told to me by elders in tea houses, and in several cases, to small snippets of "local events" revealed in the archaeological newsletters from Sardis and Turkish newspapers.

Further details regarding policies are embedded in the correspondence files and photographs of archaeologists. These include the papers of William Hepburn Buckler at Yale University, where archivist Jessica Becker provided needed guidance. Further details about Buckler's colleague Howard Crosby Butler came through the archives at Princeton University. The support and encouragement of Julia Gearhart, archivist at Princeton, became invaluable. Her time to explore Butler's letters, folders, and images that had not previously been given digitization priority at Princeton enabled me to gain access to these unpublished voices and images. Other images were discovered in the collections of the Gennadius Library of the American School of Classical Studies at Athens (ASCSA). Senior Librarian Irini Solomonidi and reference archivist and historian Eleftheria Daleziou as well as archivist Natalia Vogeikoff at ASCSA were all extremely helpful. Early images of the region at the University of Chicago, too, give us critical pictures of the region before major infrastructural changes, and Andrew Wright gave his time in sorting through the early twentieth-century images taken by Olmstead and Harris archived at the Oriental Institute.

For the mid-twentieth-century voices, the most informative sources were the personal correspondences of archaeologist George M.A. Hanfmann, archived in the Widener Library at Harvard, and those of architect A. Henry Detweiler, archived at Cornell University. Additional details were also found in the archives of the Archaeological Institute of America (AIA), the American Schools of Oriental Research (ASOR), the Sardis archives (Harvard and Cornell offices, and the excavation house), and the archives of the Museum of Fine Arts (MFA), Boston. Specific individuals in each of these places made this work possible: Kevin Mullen and Julia Grabianowski at the AIA; Andy Vaughan at ASOR; Baha Yıldırım, Elizabeth Gombosi, and Katherine Keifer at Sardis; Laura Gadbury at the MFA, Boston. In addition, the late Professor Crawford H. Greenewalt, Jr., shared his personal files and perspectives on information embedded in Hanfmann's letters as well as his personal archive of photographs and correspondence regarding the seventeenth- through early twentieth-century work at Sardis and during the Turkish War of Independence. He also shared insights regarding the initial developments at Pomza and the long-term developments in Bin Tepe. Conversations with Nancy Ramage, David Mitten, Clive Foss, Teoman Yalçınkaya, and the on-site Sardis staff were also extremely valuable. As I cross-checked as many of the events told to me and referred to in correspondences with how they were (or were not) portrayed in US, UK, and Turkish newspapers, I uncovered further networks.

Throughout this research, I also spent time in the field. I walked with farmers over their lands, discussed details over tea during rainy winter days, and spent time harvesting crops. I attended the International Fairs in Izmir focused on organic agriculture, and visited the warehouses and primary business establishments of Turkish and foreign companies. My approach to understanding more about extraction in the provinces of Izmir, Manisa, Uşak, and Istanbul followed a similar approach. I visited coal mines in Soma, nickel mines in Turgutlu, and gold mines in Manisa and Uşak. I toured marble quarries in Manisa and Izmir. Public relations officers and specialists at the mines and quarries graciously toured me around. I also attended international fairs for mining and marble held in Istanbul and Izmir.

Over the years, many colleagues and students have offered their time to have conversations with me and have invited me to present my research at their institutions, and many graciously read draft chapters. Various aspects of this research have been presented as part of the AIA Nancy Wilkie lecture series in 2013–2014 as well as in individual talks at the American Academy in Rome, the Anthropology Department at Cornell University, the Cultural Heritage Center at the University of Pennsylvania (especially discussions with Richard Leventhal and Brian Daniels), the Center for Archaeology at Stanford University, Koç

University (Research Center for Anatolian Civilizations in 2009 and Alessandra Ricci's 2016 Pubic Archaeology course in the Department of Archaeology and History of Art) in Istanbul, Mimar Sinan University in Istanbul, the Oriental Institute at the University of Chicago (and discussions with Morag Kersel at DePaul University), and Yaşar University in Izmir. Feedback from students and faculty who attended these lectures was invaluable.

I am very grateful for comments on drafts of the manuscript. Colleagues were very patient with me and took their time to read and comment on the text. These include Camille Cole, Erdem Kabadayı, Mehmet Kentel, Emmanuel Moss, Chris Roosevelt, Zeynep Türkyılmaz, Fikret Yegül, and Kathryn Lefrenz Samuels. Lynn Meskell's valuable insights and encouragement, too, have been especially wonderful. Fikret's detailed comments on aspects about Sardis, both the early nineteenth-century work of Butler and more recent endeavors, were extremely important. Kathryn and Zeynep provided their critical insights into extraction studies, and Mehmet into infrastructure studies. Chris Roosevelt graciously read and commented on the entire manuscript, and more than that, he kept me focused on the end goal. Their suggestions for clarifications and improvements in wording have made the manuscript stronger.

Over the last three years, my research assistants at Koç University have tracked down key sources. Elif Doğan carefully worked through archives about US intervention in the Gediz as well as zoning and legal files pertaining to agriculture and extraction. Zeynep Kuşdil, Zeynep Özdemir, and Levent Tokün reviewed publications and archives about the Izmir International Fair and the Kültürpark. In addition, members of CLAS were helpful in the research about water management. Elvan Cobb, especially, took part in the initiative fieldwork, as did Nedim Büyükyüksel, Nicolas Guathier, and Kyle Egerer. Ebru Kiras and Tunç Kaner, too, have spent time in local museums and touring the region with me. The communities in villages throughout the Gediz Valley, too, were gracious with their time, sharing not only oral histories but also family photos and archives.

Archival work often requires travel. Friends opened their homes to me during my visits to various cities. This was an opportunity to not only attend to my scholarship but also reconnect with people who have inspired me—and the people who reminded me to get out, to do other things. My extended family, too, made sure that this book did not consume me. They also were supportive when our immediate family moved to Turkey, and through the turbulent times of 2015 and 2016 in Istanbul, they trusted our decision to stay put.

Even still, two young men have been thrown into this study without choice, and it is to them that I owe a tremendous debt. They will have grown up primarily abroad in Istanbul, and with long stints in Izmir as well as two villages in Manisa. Their dry humor and understanding are as much a part of who they are as my husband's steadfast encouragement of my research and his unfailing support for my time to pursue it, however odd it may have seemed at the time. I've benefited tremendously from his depth of knowledge of Anatolian archaeology and the practice of it. Upon reflection, this journey began the day we witnessed a backhoe scraping away the slopes of the Sardis Necropolis as our puppy ran circles in the shadows of the Temple of Artemis.

ABBREVIATIONS

ACLS	American Council of Learned Societies
AIA	Archaeological Institute of America (US)
ASCSA	American School of Classical Studies at Athens (US)
ASOR	American Schools of Oriental Research (US)
CIA	Central Intelligence Agency, US
CIAM	International Congress of Modern Architecture
CLAS	Central Lydia Archaeological Survey
CSR	Corporate Social Responsibility
EEC	European Economic Community
EIA	Environmental Impact Assessment
EU	European Union
FDI	foreign direct investment
IBA	Important Bird and Biodiversity Area
IBRD	International Bank for Reconstruction and Development
ICIC	International Committee on Intellectual Cooperation
ICOM	International Council of Museums
ICOMOS	International Council on Monuments and Sites
IMF	International Monetary Fund
IMO	International Museum Office
IUCN	International Union for Conservation of Nature
JAMMAT	Joint American Military Mission for Aid to Turkey
KAP	Kaymakçı Archaeological Project
KTA	Knappen Tippetts Abbett Engineering Company of New York

MDG	Millennium Development Goals
MET	Metropolitan Museum of Art, New York
MFA	Museum of Fine Arts, Boston
MNC	multinational corporation
NATO	North Atlantic Treaty Organization
NEH	National Endowment for the Humanities (US)
NSF	National Science Foundation (US)
OECD	Organisation for Economic Co-operation and Development
OUV	Outstanding Universal Value
PL 480	Agricultural Trade Development and Assistance Act
PPP	Public-Private Partnership
TAMS	Tippetts, Abbett, McCarthy, and Stratton
TNC	transnational corporation
TVA	Tennessee Valley Authority (US)
UN	United Nations
UNESCO	United Nations Educational, Scientific and Cultural Organization
UNDP	United Nations Development Programme
USAID	United States Agency for International Development

Archives

AES	The Archaeological Exploration of Sardis
AHD	A. Henry Detweiler Correspondence
AIA	Archaeological Institute of America
APCA	Ahmet Piriştina City Archives and Museum, Izmir
ASES	American Society for the Excavation of Sardis
ASOR	American Schools of Oriental Research, Boston
CU	Cornell University
DAA	Department of Art and Archaeology
HCB	Howard Crosby Butler Correspondence
HU	Harvard University
JFK	John F. Kennedy Presidential Library
LCF	Le Corbusier Foundation
MA	Manuscripts and Archives
MFA	Museum of Fine Arts, Boston Archives
NA	United States National Archives
PU	Princeton University
SC	Sardis Correspondence

SM	Smithsonian Institution
TNA	Turkish National Archives
TVA	Tennessee Valley Authority Archives
UC	University of Chicago Archives
WHB	William Hepburn Buckler Correspondence
WUA	Widener University Archives
YU	Yale University Library

A Pearl in Peril

Introduction
Pillars of Policy

I woke with this marble head in my hands; it exhausts my elbow and I don't know where to put it down. It was falling into the dream as I was coming out of the dream so our life became one and it will be very difficult for it to separate again.

—"MYTHISTOREMA," BY GEORGE SEFERIS, 1935[1]

Look at beautiful Izmir, look at the golden necklace
The pearl of the Mediterranean, the leader in production
See producers and buyers exchange, let the world watch in
  astonishment
Let the crowds fill the fair, [my sweetheart] let's meet at the fair.

—"BAKIŞLARI HOVARDA, BULUŞALIM FUARDA," BY SADETTIN KAYNAK, 1939[2]

In Izmir, it is definitely Middle Eastern, with a heavy overlay of remembrances of things Greek. . . . Turkey has also been changed from one country to a completely different country because of a tremendous amount of foreign help.

—INTERVIEW WITH US DIPLOMAT WILBUR P. CHASE, 1990[3]

ON A CRISP, CLEAR June day in 2012, I ventured to Kadifekale, Izmir's high castle, with a colleague and her students. We'd come to purchase conservation supplies in the maze of Kemeraltı. I knew that the castle was slated for restoration, and that much of the slope between Kadifekale, the ancient agora below, and the Jewish and Ottoman heritage of Kemeraltı were part of ongoing gentrification programs. So, too, was I aware of the displacement of entire neighborhoods along the slopes of Kadifekale, some having been in place for close to fifty years.[4] Without deeds, more than seven thousand people living in "squatter" or "shanty" settlements (*gecekondu*) had limited legal recourse. Despite this recent tortured history, we stopped to admire the expansive view of the bay and a few of the key features in the city below: the towers of the Hilton Hotel, the Kültürpark, the Kordon, and the bustling international port. At our feet was a new public park with fresh plantings in straight rows and decorative benches. As we began our discussion with the students, an elderly man with a traditional Muslim head covering (*takke*) and a tethered goat emerged from the nearby neighborhood. In a matter of minutes the man slit the goat's throat, and blood seeped onto the stones of the park.

The histories of Izmir and Smyrna, as she was known among Europeans prior to 1922, and her countryside are not easy.[5] The past in Izmir is often described as too difficult, best forgotten. The municipality hopes that gentrification will draw tourists, Turkish and foreign, into the city proper. Urban renewal, however, is a challenge in this city. Many continue to identify with the profound sense of loss and abandonment, as portrayed by George Seferis' 1935 poem *Mythistorema*.[6] This is not merely nostalgia, nor is it a mythologizing of historical events where reality has been radically blurred. While the 1922 "Asia Minor Catastrophe" has been infused in US diplomatic memory as the period when the West lost the doorstep of Anatolia, it has come to embody the deep misunderstandings by and mistrust of foreign powers about the complexities of western Turkey.[7]

To be sure, there are also narratives that speak to the revival of the city after the fires of 1922, the reawakening of Izmir, and the importance of this region on the global stage. In 1939 Sadettin Kaynak's lyrics, quoted at the outset, marked the economic success of Izmir's International Fair. This city, the "golden necklace," symbolized "the pearl of the Mediterranean, the leader in production." The juxtaposition between US diplomatic wistfulness for what could have been and Sadettin's account of Izmir's revival demonstrated the vexed position of this region in the 1930s and foreshadowed a future of political intrigue and economic competition.

Today Izmir is Turkey's third largest metropolitan area. During the early twentieth century a cosmopolitan lifestyle here rivaled those of other eastern Mediterranean ports: Alexandria, Beirut, and Athens. Muslim communities reportedly referred to the city as "*gavur*" (infidel) Izmir.[8] Despite the city's reputation as among the most liberal in Turkey, minority groups have slowly dwindled over the years. The Jewish exodus in the late 1940s and early 1950s has been followed with more punctuated periods of departure, most recently Spain's June 2015 invitation for

Iberian Sephardic Jews expelled in the fifteenth century to return.[9] Other groups, such as Greeks and Armenians, and the Levantine communities, too, have seen their numbers diminish. Even so, the city cultivates its liberal image and its steadfast alliance with the Kemalist Republican People's Party (CHP, *Cumhuriyet Halk Partisi*), the counter party to the Justice and Development Party (AKP, *Adalet ve Kalkınma Partisi*).

This tension seeps into Izmir's hinterland, especially along the Gediz River valley and into the regions of Manisa and Uşak. Agricultural fields, dams, stone and gravel quarries, and gold mines supply Izmir with food, water, electricity, and materials for export. Unlike liberal Izmir, however, the politics of the hinterland reflect shades of grey with pockets of support not only for CHP and AKP, but also for the Nationalist Movement Party (MHP, *Milliyetçi Hareket Partisi*). The mosaic of politics creates an eclectic environment that intermingles the diversity embedded in the past with new visions for the future. Through this lens, Izmir and her surrounding regions may never be what foreigners planned for her; nevertheless, she is comfortable in her own skin, a city and a countryside with a set of subtle, cosmopolitan social codes open to the outside world, yet also weary of it given the collective memory of trauma and distrust.

Over the last decade, the majority of foreign representatives (for example, John Deere (US), Pepsi (US), International Harvester (US), Bosch (German), and Gaz Group (Russian)) who have attended the international fairs held in Izmir's Kültürpark were unaware of the tortured histories on which they stood: the Armenian neighborhoods between Alsancak and Basmane had been scorched in the fires of September 1922.[10] Those were the final hours of the Turkish War of Independence, when the military forces led by Mustafa Kemal drove Greek military forces from Asia Minor. The infamous flames destroyed much of Smyrna, just as other fires had razed towns and villages inland during the course of the Greek military occupation and retreat (May 15, 1919–September 9, 1922). The cultural landscapes of Izmir are imbued with this date, such as the 9th of September University and the Kültürpark's 9th of September Gate.[11]

Eighty years later, sentiments of sorrow resurface each September in the annual celebrations of liberation from Greek forces in the Gediz Valley, as each town commemorates its respective victory day. The festivities, supported by local municipalities and governorships, culminate in Izmir on the 9th of September. Enormous Turkish flags and banners of Turkey's first president, Mustafa Kemal Atatürk, wrap buildings. At 9:00 a.m. canons fire into the bay, where further out battleships take position.

At midday military planes buzz the city. Evening bands play until after midnight, when fireworks light the night sky. Between the canons and the fireworks, parades of schoolchildren, clubs, sports teams, police, and Jandarma march through the streets and along the waterfront of the Kordon (an open green space and public park along the Izmir bay) toward Republican Square.

Here citizens gather to hear officials speak. They stand in the shadow of the 1932 bronze equestrian statue of Atatürk by Italian sculptor Pietro Canonica (Figure I.1). Atatürk's right arm is raised and his forefinger points toward the sea. His command still resonates: "Soldiers, your goal is the Mediterranean." The performance of this image forms part of a countrywide mental map and a national consciousness that solidifies Izmir's Kordon in the official historiography of the Turkish War of Independence.[12] This is the heritage that many find too painful to discuss, yet compulsory to commemorate.

FIGURE I.1 Atatürk sculpture by Italian sculptor Canonica, Izmir. Photo by H. Basak. Courtesy of Creative Commons.

The millions of national and foreign tourists who have flocked to the shores of the Aegean beaches and traversed the Classical ruins of Ephesus, Aphrodisias, and Halicarnassus—among others—that dot the countryside, however, are mostly ignorant of the decades of deceptive diplomacy that have shaped the contemporary Republic of Turkey, and Izmir and her hinterland specifically. Nonetheless, it is precisely the well-trodden Kültürpark, the tourist zones of archaeological sites, infrastructure projects for roads and water, and the newly gentrified neighborhoods that give us a glimpse of foreign intervention in the final decades of the Ottoman Empire and subsequently after the rise of the Republic. For over a century, archaeology and preservation have been pawns caught in political moments that can be traced to careless diplomatic practices, poorly spent development dollars, heavy-handed ideological frameworks, and the evolving strategies of security.

Intersections between the nation-state, heritage, and archaeology have been emboldened by recent political platforms, foreign and national, that merge ethnicity and religion into narrow categories that are often unrealistic for daily life. This practice creates an untenable situation whereby people, countries, and regions are forced into categories defined by authorities and state actors, rather than a flexible and fluid process that enables self-identification and associated place-making. Staged destruction of heritage by insurgent groups in Afghanistan, Iraq, Syria, Yemen, and Libya has given UNESCO's World Heritage Center renewed visibility. These shock-and-awe events have also been met with rallying calls to ramp up global preservation efforts.[13] This process has in turn resulted in greater prestige for World Heritage nominations and renewed focus on the rights of minority groups and heritage governance in the name of "the global good." The practice in reality, however, is a different matter. Policies that support mobility and sanctuary, particularly rhetoric that emphasizes "safe havens" and "safe passage" for diaspora and refugee groups, often appropriate heritage for political ends.

At the center of these global debates lies modern-day Turkey. There is steady media coverage regarding the future of Turkey vis-à-vis stability in the Middle East, Europe, and Central Asia as it pertains to political structures, military agendas, and civil society. Politicians in Europe and the United States still struggle with the "Eastern Question" and Turkey's geopolitical position.[14] The years between 2015 and 2018 have been especially turbulent. Flooded by refugees from Syria and Iraq, Turkey has become

Europe's indispensable buffer zone.[15] New networks for smuggling—from people to materials—across the Iraqi and Syrian borders into Turkey have fostered illicit networks in all spheres, including the trade in antiquities.[16] So, too, have global conversations and events brought contested spheres of heritage to light in Turkey.

During the 40th World Heritage session in Istanbul in 2016, concerns about places of minority heritage in Turkey were whispered over teas and coffees. The topic, however, was not openly debated. Still, the sentiments were implicit during the deliberations and carefully crafted political speeches in response to the inscription of the Archaeological Site of Ani, the former Armenian capital of this medieval kingdom, which occurred on the afternoon of July 14, 2016. Less than a month earlier, the German Parliament had voted to recognize the 1915–1916 massacres of Armenians as a targeted effort; the Turkish government did not agree with this vote.[17] In retaliation, Turkey recalled its ambassador and withheld 2016 German permits for archaeological research. On the floor of the World Heritage session, one would have never known about this and other related tensions regarding Armenian heritage and history. In a show of support for future collaboration, Turkish and Armenian diplomats called for shared dialogue and transnational partnerships. A great sense of hope pervaded the room. The attempted military coup in Turkey that evening dampened this glow—the delegates huddled in their hotel rooms as planes took to the skies and tanks blocked the bridges over the Bosphorus. The meetings were suspended until late October 2016 in Paris, yet civil society groups in Turkey looked to the Ani inscription for future promise of increased dialogue.[18]

In video footage from the World Heritage session, Osman Kavala is shown taking pictures of Ambassador Vahram Kazhoyan, Secretary General of the Armenian National Commission for UNESCO, during his speech. Kavala is the executive director of Anadolu Kültür known for cultural heritage activities (including at Ani). A year later, on October 18, 2017, Kavala was detained by police at Istanbul's Atatürk International Airport, and in the early morning hours of November 2nd was formally arrested.[19] The legacy of the Kavala family is particularly poignant: they were relocated from northern Greece (Kavala) in the population exchanges following the Turkish War of Independence.

The Turkish–Greek population exchanges of 1923 focused first and foremost on religious identity, whereby Muslim families were made to leave Greek territory and emigrate to Turkey and vice versa.[20] Foreshadowing of these future tensions was clear during the 1919 Paris Peace negotiations. At these sessions voices from Turkey were not heard, even symbolically; others spoke for them. In the proposed Treaty of Sèvres, Articles 140–151 sought protection of minorities, and along similar lines Article 230 sought resolution and restitution of personal property and persecution of crimes committed against minority groups during World War I in Ottoman territories; yet on the other hand, Part 9 (Economic Clauses), Part 10 (Aerial Navigation), and Part 11 (Ports, Waterways, and Railways) detailed the extent to which Turkey was expected to abide by Allied ambitions for foreign concessions and business arrangements.

Sèvres also offered a template for the Allies to push their vision for the management of antiquity. This included the desired territories detailed in Article 66. The intended boundary for Greek occupation in western Asia Minor would have subsumed the Classical sites of Ephesus, Sardis, Philadelphia, and Pergamon—and the respective fertile river valleys and mountain ranges with significant resources ripe for development (e.g., agriculture and extraction). The underlying agenda of the Greek campaign was to revisit and pursue the *Megali Idea*—a dream of sovereignty over zones with claims to Hellenic history.[21] Proposed in the mid-nineteenth century by Greek Prime Minister Ioannis Kolettis and King Otto, the general concept figured prominently into discussions of 1919 in Paris. British Prime Minister Lloyd George and US President Woodrow Wilson agreed not to thwart Greek intentions.[22] Among the most ardent advocates was career diplomat George Horton, a fervent Philhellene who had served as US Consul in Smyrna (1911–1917, 1919–1922), US Consul in Athens (1893–1898, 1905–1906), and US Consul in Salonika (1910–1911).[23]

Classical archaeology played a significant role in establishing jurisdiction in Smyrna's hinterland, as noted in Article 421 of Sèvres:

> The Turkish Government will, within twelve months from the coming into force of the present Treaty, abrogate the existing law of antiquities and take the necessary steps to enact a new law of antiquities which will be based on the rules contained in the Annex hereto, and must be submitted to the Financial Commission for approval before being submitted to the Turkish Parliament. The Turkish Government undertakes to ensure the execution of this law on a basis of perfect equality between all nations.

Both Articles 66 and 421 sought to empower an "International Commission for Antiquities for the administration of historic monuments in Turkish lands, acting as the mandatory of the League of Nations."[24] For the most

FIGURE I.2 Map of Treaty of Sèvres and featuring E. Venizelos. Courtesy of the American School of Classical Studies at Athens, Gennadius Library.

part, news about discussions in Paris focused primarily on the urban hub of Smyrna; considerations about the vast territory encompassed under the proposed area were less present. And yet, it was the hinterland that many sought: access to mines, fertile river valleys, infrastructure (train lines), and archaeological sites of interest. In Figure I.2, Greek Prime Minister

Eleftherios Venizelos is shown above a map that details the territory of Smyrna placed under Greek jurisdiction.

The Treaty of Sèvres was never legally promulgated, and thus the desires of the Allies were never fully realized. The Treaty of Lausanne (signed July 24, 1923) did not include provisions for the management of archaeological sites nor objects. Nonetheless, the correspondences of those who participated in the discussions in Paris and activities in Asia Minor at the time give a compelling history of how archaeologists participated in the art of diplomatic practice in a symbolic city and a coveted countryside. We're now in a position to reflect on the consequences of these choices over the last century.

I approach heritage studies from the perspective of development and ask how and why heritage becomes entangled by heritage statecraft, corporate contracts, and, more broadly, practices of social engineering.[25] I'm especially interested in patterns of top-down paternalism displayed when authorities define and enforce social codes of conduct for communities. This is common in nationalist agendas and is woven into the fabric of urban planning and education. It is also a common rationale for assistance programs by "developed" nations for "underdeveloped" nations, the former offering moral authority through a set of standardized ideas and implementation procedures, usually linked to economics, politics, and persuasion.

In 1919 archaeologists sought an international commission to oversee archaeology in Turkey. Today, heritage statecraft includes concentrated and sustained efforts to push forward with dossier preparation for World Heritage nominations for sites with Outstanding Universal Value (OUV). The United Nations pledges a transparent, untainted review process. That scholarship reveals the opposite should not be surprising.[26] For centuries, nations have manipulated history for their own purposes.[27] In the early 1930s, for example, Turkey drew extensively, yet selectively, from archaeology when constructing her collective identity and narrative of self-protection under the new Republic.[28] With each revisionist approach, whether in Turkey or elsewhere, the past can be carefully reconfigured, reevaluated, and understood anew with the sweep of a single event, whether due to social discourse, political pressure, military intervention, or natural disaster.

Studies regarding World Heritage in Turkey have produced an abundance of literature.[29] In fact, many sites owe their current performance on the global stage to foreign assistance programs that occurred decades before the 1972 World Heritage Convention. These histories of "aid," however, rarely grace the pages of World Heritage documents. At best, those who funded the projects receive a few sentences. Without a deeper look into the initial policies and research that framed knowledge about the respective sites and the agendas of those who funded the projects, many analyses of World Heritage areas remain floating in time. A critical analysis of the legacies of promise, the unheard voices, and hidden histories allow for a more nuanced perspective of how heritage comes to be defined, challenged, erased, forgotten, and resurrected. These issues raise their heads now through the interests of international and national foundations, criteria for competitive grants, and increasingly Corporate Social Responsibility (CSR) programs that are branded as a moral face of neoliberalism.[30]

In the chapters that follow, I trace and analyze the operations of diplomatic and archaeological practice and actions of development that have sought to create and realign territories of heritage over the last century. In each chapter I consider the major policies, both national and international, that helped to shape the development sector in question, and in turn the landscapes of the middle Gediz Valley, the zone that feeds and waters the city of Izmir and beyond the Mediterranean Sea.

Chapter Overviews

At the heart of the analysis that follows is the issue of policy. Often presented as a moral framework that is softer than law, policy in general offers greater potential for dialogue. Yet, precisely because policy can be flexible, it may assist in shifting legal frameworks more quickly than we might think, especially when we're considering a policy change that will usher in major revenue to state coffers and the pockets of the private sector. Under these conditions, it is possible for policies in one sphere to change much more quickly than in another. Scholarship has shown just how poorly we currently understand transnational approaches to development and social engineering.[31] Throughout the course of this study, I examine punctuated periods of policy changes that occur during the realignment of international territories, nation building, neoliberalism, and most recently transnational corporatism.

Among the fastest-moving areas of new policy in Turkey is the robust and aggressive industry of extraction, and it is for this reason that I begin

Chapter 1 with this topic. In addition to coal, natural stone and gold are the top resources extracted from the hinterlands of Izmir, most often made possible through transnational partnerships and CSR programs. Hillsides are shaved and penetrated by the aggressive marble and limestone quarrying industry to feed the ever-growing appetite for cement and designer stone markets. Yet, local communities aren't always pleased by these operations. I illustrate one case study where the community pushed back using archaeological zoning.

For the growing industry of technology and the ever-shaky financial systems, gold is the metal of choice. Located 180 kilometers east of Izmir is the largest gold mine in Europe. At the Kışladağ mine in the province of Uşak, Tüprag and its Canadian parent company, the Eldorado Gold Corporation, oversee operations. Tüprag's successful CSR programs have engendered a culture of acceptance for large-scale extraction because social mobility in the rural areas of Uşak has been considerable. Smaller-scale gold mines, such as Pomza Export in the province of Manisa, have also been able to leverage the favorable climate for extraction and to promote CSR. This power has allowed Pomza to operate within the archaeological territories of Sardis, listed in 2013 on the World Heritage tentative list *The Ancient City of Sardis and the Lydian Tumuli of Bin Tepe*.

The mega-projects and extraction forces of the twenty-first century in many ways mirror the lust for Anatolia at the turn of the nineteenth century. It is for this reason that I move from an analysis of early twenty-first-century conditions in Chapter 1 to those of the early twentieth century, the subject of Chapter 2. From this point, I follow a chronological progression from the late nineteenth century to the present. In each chapter, I demonstrate how the respective policies and laws that sought to preserve and protect areas of heritage in the middle Gediz Valley confronted those that favored economic development, particularly programs that privileged a "multiplier effect" and a "redoubling" investment model. In this way, the countryside has been and continues to be engulfed by the needs of a major urban core, but also economic potential driven by export markets. At the current pace of development, the failure to harmonize cultural policies with neoliberal economics threatens heritage at an unprecedented scale, despite decades of policies and laws that have sought to do just this.

Chapter 2 explores the place of early twentieth-century Smyrna in the context of the 1919 Paris Peace Conference, the mission of the US Archaeological Institute of America (AIA), and the archaeological expedition to Sardis of Princeton University. The unofficial advisors of the AIA were Howard Crosby Butler and William Hepburn Buckler. Both were

Princeton men and both had been integral to the first American archaeological expedition to Sardis, Butler as excavation director and Buckler as epigrapher, financier, and diplomat. Through these two men the AIA collaborated closely with the US Congress to determine avenues for top-down US cultural diplomacy abroad. A thick web of personal and professional entanglements arose between 1910 and 1919 regarding the vision of the AIA for the management of antiquities in Asia Minor.

The initiatives of 1919 in Paris for an international commission to oversee archaeology in the Mediterranean, Asia Minor, and the Near East failed, but new programs were formed. In the cultural dimension, these included pledges of mutual understanding and shared dialogue through the League of Nations' International Committee on Intellectual Cooperation (ICIC). The promise of exchanging information to mitigate future conflict was attractive to many participating members. The ICIC's goal followed a positivist approach, seeking a science of universal moral authority.[32] Under the ICIC, the International Museum Office (IMO) sponsored conferences on preservation and archaeological practice. The 1931 International Congress of Architects and Technicians of Historic Monuments was a major starting point, known as the Athens Charter of 1931. It offered international standards for conservation and restoration, especially methods of anastylosis.

Another Athens Charter was reached in 1933 by the International Congress of Modern Architecture (Congrès Internationaux d'Architecture Moderne, CIAM). CIAM explored not what the past was, nor how to recreate it, but rather what tomorrow could be, and what the future should be. The group sought to resolve urban issues through new design. Members grappled with "heritage zones," seeing archaeological sites not as objects, but as an embodiment of values. New cities were designed with the hope that architecture and urban planning could stimulate social cohesion. Among the most prominent figures for CIAM was Le Corbusier, who in 1939 was hired to prepare plans for Izmir.

Chapter 3 tackles the ideological importance of CIAM in a larger conversation about water diplomacy and the tremendous influence of the US Tennessee Valley Authority (TVA) in the Gediz Valley. The Aegean-TVA hydro-heritage landscapes symbolize the ideological tensions that arose between 1930 and 1960, a period that saw the demise of the League of Nations and the début of the United Nations, and with it the establishment of the United Nations Educational, Scientific and Cultural Organization (UNESCO). The TVA model had significant impact on Julian S. Huxley, the first Director-General of UNESCO. Huxley's 1943 book *TVA, Adventure in Planning* captured his interest in design and technology and, more significantly, the potential to integrate hydroelectric power, irrigation, mechanized agriculture, and social engineering. He admired TVA's power of persuasion. Leadership from TVA joined UNESCO projects, exploring the river valleys in Turkey, Syria, Jordan, Iraq, and Egypt, where the Cold War theater of the High Aswan Dam in Egypt took on a cultural dimension in the salvage operations at the famed Abu Simbel complex and other monuments in Nubia.

Chapter 4 considers US intensions and long-term commitments to international heritage in the shadows of the Nubian campaign. I demonstrate that while the public perception of rescuing heritage at Abu Simbel portrayed a solid ideological and financial commitment, including endorsement from US President Kennedy, US Congress had no intention of a holistic, sustained global initiative. The evidence is embedded in the correspondences of A. Henry Detweiler (Cornell Professor, President of ASOR, and Associate Director of Sardis), George Hanfmann (Harvard Professor and Director of Sardis), and Jotham Johnson (Professor at New York University, President of the AIA, and Director of Aphrodisias), and men in Washington, D.C.: Lucius (Luke) Battle at the US Department of State and William Warner at the Smithsonian Institution. Sardis hoped to build on the success and momentum generated by Abu Simbel, yet the US Congress had no plans for future funding of heritage projects. This case study demonstrates the irony of policies and perceptions regarding US pledges and the domino effect from high-profile projects with exorbitant costs.

In Chapter 5 I bring the issues explored in Chapters 3 and 4 together. The royal burial area of ancient Sardis, known as Bin Tepe, is composed of the symbolic landscape of earthen burial mounds, or *tumuli*, that stretch 72 square kilometers (ca. 45 square miles) across a limestone ridge on the shores of Lake Marmara, just beyond the current coarse of the Gediz River. The tumulus of the celebrated Lydian king Alyattes is located here, and at one time there were likely as many as 140 tumuli in Bin Tepe. Currently the backhoe and the tractor threaten this landscape, the former driven by the illicit trade in antiquities, and the latter fueled by a lack of harmonization of heritage policies and agricultural development of organic olives and grapes, technically illegal yet the heart and soul of social mobility for local communities.

The topics presented in the chapters of this book span a considerable period of time from the late nineteenth century through 2018. The case studies and narratives are intended to demonstrate the political consequences to be expected when heritage is ignored during crisis, whether military or economic, or over the course of expansion, most often associated with infrastructure, agriculture, and extraction. This approach allows the reader to engage in a historical analysis from multiple perspectives that recognizes the impact of social difference and political agendas at archaeological sites, themselves subject to academic pursuits laced with specific ideologies. This exposé on the bureaucracies of heritage demonstrates the need to move beyond the fixed, known locations, such as the famed home of Croesus at Sardis, to understand that these "known" places are portals into narratives of social engineering that challenge our sensibilities of change over time and how history is deployed in defining the present for the future.

CHAPTER 1 | Alternative Futures

The development of a nation is dependent on the evaluation and exploitation of its subterranean resources.

—MUSTAFA KEMAL ATATÜRK[1]

Minerals are under the sovereignty and disposition of the State, they are not under the proprietorship of the landowners of the land where they exist.

—Article 4, Turkish Mining Law No. 3213, as amended, No. 5177

VISIONS OF TOURING THE seven churches of Revelation or walking in the footsteps of Alexander the Great may attract foreign tourists to the landscapes of Anatolia. Yet, it is the lure of natural stone and the promise of striking gold that entices foreign investors. Areas once quarried for marble temples and mined for the world's first gold coins have also turned handsome profits for twenty-first-century transnational companies. The footprint is prominent and global, such as the "Turkish Burdur Beige Stone" that covers the floors of Terminal 1 at John F. Kennedy International Airport in New York City.[2] In March of 2017, among the towering green stone columns, sleek escalators, and walkways encased in glass at the international Izmir Adnan Menderes Airport, billboards promoted not only the marble façade of the Library of Celsus at Ephesus, but also Izmir's 23rd International Natural Stone and Technologies Fair, MARBLE. This is among the largest expos in the world for this specific market. Slogans for MARBLE 23 read, "Experience our new beige select: moon cream and vanilla silver." On other billboards, Tüprag advertised their commitment to "Turkey's golden future," including promotion for their gold mine in Uşak, Europe's largest.[3] An image of a stylized green plant emerging from a gold bar in the upper right-hand corner of the billboard evoked the pine tree saplings that representatives of coal mines handed out at the 2016 Mining Fair in Istanbul.

Caught in the crossfire, archaeology finds itself navigating new territories dictated by these multimillion-dollar deals and their robust revenue streams. My analysis of heritage and extraction in western Turkey in this chapter stems from fifteen years of ethnographic fieldwork in the regions of Izmir, Manisa, and Uşak. During this period I spent two months each year working on archaeological projects based in small villages as well as two years in residence, one in Manisa (2000–2001) and another in Izmir (2010–2011). Whether during the summer months of archaeological fieldwork or the years in residence, the projects that I was involved with (Archaeological Exploration of Sardis, Central Lydia Archaeological Survey (CLAS), and the Kaymakçı Archaeological Project [KAP]) were often consulted by regional museums, the Izmir Second Protection Council, the Ministry of Culture and Tourism in Ankara, and consultancy agencies responsible for environmental and cultural impact assessments for new infrastructure projects. As archaeologists, we were asked to provide data and to prepare maps in response to the development of highways, quarries, and mines.[4]

The kings of extraction, their technologies, and their markets (e.g., Austria, Australia, Canada, China, Germany, India, the United Kingdom, the United States, and others) are quickly transforming global landscapes in close proximity to heritage sites. In the 1990s two advisory bodies of the World Heritage Center, the International Union for the Conservation of Nature (IUCN) and the International Council on Monuments and Sites (ICOMOS), gave guidelines for working with the extraction industry. Two decades later, a 2015 report summarized extraction in the vicinity of World Heritage sites: 18 percent of fully inscribed (i.e., not including the tentative list) properties were affected by mining, oil, and gas activities, and of these 84 percent were natural or mixed properties.[5] These statistics should raise global alarm, and yet they have not. The US markets for white marble countertops and the London Gold Fix forge full-force ahead. Erasure of the natural and cultural places slated for protection in perpetuity has become the new normal.

This tension comes at a time when archaeological heritage is fiercely contested in Turkey. Demands for repatriation and ongoing legal cases are rampant. Diplomatic blunders have resulted in archaeological research permit suspension (e.g., German permits for Hattuşa and Pergamon) and even the threat of permit revocation (e.g., the Austrian permit for Ephesus).

These political moments over heritage fracture attempts at cultural diplomacy and holistic sustainable development. On the one hand, there is a global outcry against targeted violence to heritage in war zones; on the other, the very countries (e.g., Australia, Britain, Canada, Germany, and the United States) mourning the destruction of heritage in Syria, Iraq, and Yemen stay silent on extraction, despite their place in this corporate structure.

My analysis of archaeology as arbiter begins with two different narratives from the Gediz Valley in western Turkey. In the first case, the community of Hacıveliler leveraged the archaeological zones of Kaymakçı and Bin Tepe as well as the ecological area of the Marmara Lake Basin to preserve traditional lifeways threatened by a marble quarry. In contrast, the Pomza Export gold mine trumped the heritage of Sardis. The very material (gold) that contributed to the rise of the Lydian Empire has also been the material that has contributed to the erasure of large portions of her core historic landscape. In this way, the materials (marble, gravel, sand, and gold) that enable modern-day Turkey to pursue aggressive development also destroy the heritage that has for centuries defined her.

Marble Madness

During the nights of March 2015, the women of Hacıveliler slept in the cold on a dirt mountain road between their current village and that of their ancestors, nestled among the nearby foothills. During the days, they held colored poster paper with hand-written statements: "For our future, no to stone quarries." "Stone quarries are quarries of sorrow."[6] Their protest was against a planned marble quarry to be opened in land that had supported their pastoral economies for generations. Their road vigil blocked the machinery from ascending the mountain to extract samples for MARBLE 23. Their actions were not alone. These women followed the practices of other groups in Turkey who had also demonstrated to raise awareness about and against extraction.[7]

The Salihli Environment and Social Solidarity Association, the Turgutlu Environment and Culture Platform, the Salihli United June Movement, the Salihli Eğitim-Sen, and the Chamber of Chemical Engineers in the Aegean Region supported the resistance. The community of Hacıveliler (population ca. 200) was unwavering in their support for, in their words, "Our strong women. They will save our village." The territory under question was zoned as public land (*mera*) and forest (*orman*) and had been the primary grazing grounds for livestock (sheep and goats), which in turn provided significant economic returns for this small village. Villagers took issue with the lack of transparency and consultation, as the newspapers reported: "Without our consent, they will have difficulty passing! We do not give permission."[8]

Not only did villagers wish to continue their pastoral lifeways, but they also had reason to fear extraction. Hacıveliler is in the province of Manisa, western Turkey. During the negotiations of March 2015, the Soma mine disaster of May 13, 2014, was still a very raw memory. Described by many as the worst mining catastrophe in Turkish history, this tragedy shook the foundations of the Turkish coal mining industry. It also gripped the global news outlets, including the *New York Times*, which ran stories in May 2014, November 2014, and again in June 2015.[9] Over 700 men were inside the mine, some 400 meters below the earth's surface, when a fire broke out. Once rescue efforts concluded, at least 301 had perished and 80 were injured.[10] Sons and fathers of relatives and friends in the nearby center of Gölmarmara were among those who lost their lives.

What is more, the community had witnessed the destruction by marble and limestone quarries to the once pristine mountain slopes of nearby Akhisar. The open hillsides surrounding the city of Izmir, too, are evidence of this industry's aggressive approach. Residents dreaded the trucks, the noise, and, as they described it, "the heavy air" from the clouds of dust. When considering litigation, they also recalled tensions and overturned decisions concerning the gold mines in Bergama (Eurogold, an Australian and German venture, at Ovacık, Çamböy, and Narlıca), Uşak (Eldorado [Canadian] and Tüprag [Turkish] venture at Kışladağ in Eşme), and Sart (Pomza Export [Turkish]). Regardless of the hurdles in front of them, the community contested the quarry. When the court ruled in favor of the marble company, the community asked how archaeological zoning could serve their interests.

In 2011 a recently discovered second-millennium-BCE citadel known locally as Kaymakçı, once a Hittite vassal kingdom and peer of Troy,[11] had been designated under the highest degree of archaeological zoning for Turkey. The area abutted the pastoral and ancestral landscapes of Hacıveliler that were registered as forest. Adjacent zones carried badges of international significance (see Chapter 5, this volume), and thus strengthened the argument. To the immediate southeast sits the archaeological landscape of Bin Tepe, which as of April 2013 had been, with the archaeological site of ancient Sardis, placed on the Turkish tentative list for UNESCO World Heritage. To the immediate east stretches Lake

Marmara, which between 2008 and 2009 had been designated a Turkish wetland protection zone and listed as an Important Bird and Biodiversity Area (IBA) by the IUCN. The future quarry zone would have been in the buffer zone for Bin Tepe and the watershed of the Lake Marmara Basin.

In addition, submissions to the Izmir Second Protection Council for protection of other recently discovered archaeological sites in the Basin had either just been approved or were pending.[12] To verify archaeological evidence in the area of the proposed quarry, the Izmir Protection Council and the Manisa Museum of Ethnography and Archaeology requested further archaeological survey. The results revealed evidence of pre-industrial quarry activity on the faces of quarries slated to be opened, or re-opened as the evidence now indicated.[13] Additionally, extensive evidence was documented for ritual features with clear links to central Anatolia and Hittite traditions as well as sacred connections to the Lake.[14]

In July 2015 Turkish courts ruled in favor of new archaeological zoning. Plans for the quarry were stopped, and the Hacıveliler community resumed their agro-pastoral activities. There remains, however, the threat of future development, especially given the unprecedented and expanding global demand for natural stone and new quarries in the region, as shown in Figure 1.1 and the annual market at MARBLE.

FIGURE 1.1 Marble quarry in the province of Manisa with tumulus in foreground. Courtesy of the author.

Throughout the 1920s and the 1930s the Izmir International Fair was held in September each year. When the Kültürpark opened in 1937, it represented the new face of Turkey and a central venue for international fairs. From the 1940s through the 1960s, the fairs became a hotbed for political intrigue and Cold War diplomacy (see Chapter 3, this volume). In 2015 Izmir welcomed the opening of a new, enormous fairground (near the Menderes airport and Freeport zone) with extensive interior and exterior spaces for displaying blocks of marble, massive machines, and three floors of interior exposition space for corporate exhibitions gleaming with polished stone. Nestled among the indoor exhibitions are booths for consulting agencies, the Turkish General Directorate of Mineral Research and Exploration, and a lecture hall for academics and researchers.

Designers and their clients in the United States, China, and India know little about the impact of the global lust for marble on small, rural Turkish villages, such as Hacıveliler. International stylistic whims dictate where and when quarries open on Turkish soil. I learned that while American clients may covet unblemished white dolomite for their kitchen floors and countertops, the Chinese wish to clad walls and floors of restaurants and residences in variegated yellow and grey onyx, often backlit to create the appearance of fire and ice, especially attractive for night clubs.

Turkish markets, too, favor stone as a primary decorative product. From the fortified US Consulate and the Ottoman Archives in Istanbul to the magnificent new provincial museum in Şanlıurfa, stone communicates prestige. These examples of opulence and luxury set against the backdrop of industrial quarries and factories exemplify the international scale of the stone industry for Turkey, which fills a significant part of the global market. According to Turkish government statistics for 2016, natural stone was among the leading exports from Turkey.

At MARBLE 23 (March 2017) the juxtaposition between floor-to-ceiling examples of polished stone on one wall and the images of extraction zones (i.e., the quarries themselves) on the abutting wall were particularly striking. The enormous images celebrated large gashes in the landscape and extensive spoils cascading down once-pristine mountain slopes. These are scenes of horror to resolute environmentalists, but of delight to those seeking stone and profits. Corporate folks in suits and heels hob-knobbed in the white leather chairs alongside glass tables adorned with savories and sweets, including tile-shaped cookies covered with a marbled glaze. Catchy advertising titles for marble varieties, such as "sweet cream" or "black galaxy," were backdrops to the performance of corporatism. In this context, the herder of Hacıveliler seeking to protect her livelihood has some fierce competition. And she has reason for continued concern: a newly formed Turkish-Chinese partnership has its sights set on neutral, warm tones, such as "Ottoman pearl" and "Honey Onyx," from nearby Gölmarmara and Akhisar.

Gold Glory

The steady growth in the extraction of stone is matched by the expansion of the gold industry in Turkey. Uncertainty in global markets over the last two decades has led to a greater reliance on the gold standard to preserve financial stability. Gold is also a preferred metal for new technologies. The World Gold Council reports that gold is the "backbone of the internet" and critical in new technologies for health and space exploration. Global currency markets benchmark against the "Gold Fix"—the price of gold set twice daily (10:30 and 15:00 GMT) by a small group of banks in USD, GBP, and Euros. Research shows that it is not only these major banks that dictate the price of gold, but also in-country exchanges (e.g., the Istanbul Gold Exchange, opened in 1995) and on-site activities at mines.[15] Turkish companies (e.g., Tüprag, Lidya Madencilik Sanayi ve Ticaret A.S., Anagold Madencilik Sanayi ve Ticaret A.S.) and their international partners (e.g., Britain's Stratex International; Canada's Alacer, Alamos Gold, Aldridge Minerals, and Eldorado Gold; Australia's Ariana Resources; the American Mediterranean Resources; the Channel Islands' Mariana Resources) drive the gold industry in Turkey. Since the 2001 bull market, Turkey has promoted favorable legal frameworks, streamlined bureaucracy, and endorsed mining as a high-priority strategic, national initiative.[16]

Unlike the proposed Hacıveliler marble quarry, gold mining has trumped archaeological heritage in the middle Gediz Valley. The most surprising is that of Pomza Export, operating within the ancient environs of Sardis, as shown in Figure 1.2. Listed on Turkey's World Heritage tentative list in 2013, an aspect of the claim to Outstanding Universal Value for the ancient city of Sardis and her hinterland is the heritage of gold:

> The city is dominated by the ancient acropolis, which rises 300 meters above the surrounding plain, "the strongest place in the world" according to Polybius (9.20.12). The Pactolus stream (modern Sart Çayı) runs by the base of the citadel, and was famous in antiquity for its alluvial gold, making the Iron Age Lydians the wealthiest people in the world. . . . [There is a]

FIGURE 1.2 The foothills of the Tmolus mountain range, including the Sardis Acropolis, Sardis Necropolis, and Pomza's operations. Courtesy of the author.

> Lydian industrial sector outside the city walls, which preserves the earliest evidence in the world for the refining of electrum into pure gold and silver.[17]

Despite this legacy, on a late fall afternoon in 2000 from the Temple of Artemis of Sardis, I watched a backhoe move across the upper tree line of the Necropolis ridge. The boom and arm were scraping earth from the face of the mountain. The machine had emerged from a deep cut through the eastern ridge, and below it the initial infrastructure for the settling tanks and offices was underway.

Inquiries over the next year-and-a-half revealed that Pomza Export Mining Industry and Trade, Inc. had secured a concession for borax. Not long after, their website read, "We begin where the Lydians left off." The mission for gold was clear. A few newspaper articles voiced discontent, yet overall there was little press. In subsequent years, punctuated moments of tension arose regarding fears of toxicity and environmental degradation. Locals lamented the future of their village and claimed in 2003 that olive trees were dying. Collective memory associated dead trees with the devastation of fruit trees, particularly pears, which had supposedly occurred after the catastrophic Chernobyl accident in April 1986. In contrast, there was little community concern for the archaeology of Sardis because in the minds of locals the Harvard-Cornell directorship of the Archaeological Exploration of Sardis and the Manisa Museum of Ethnography and Archaeology controlled the site, and over the decades this alliance had held significant power to thwart development that threatened the integrity of the site. But, in this case, gold outmaneuvered archaeology. When the mine went forward, locals dreaded not the erasure of history, but rather the cessation of the Expedition and with it fewer options for employment.

Summary

These two case studies illustrate the coalescence of Turkish tangible and intangible heritage confronting the international demand for marble and gold. In the remaining sections of this chapter, I place these two industries in a wider context, drawing out the entanglement of politics and strategic diplomacy. I also unpack some of the malleable interpretations regarding land management, productivity of forests, and perceptions of "a high quality of life" now enshrined in Corporate Social Responsibility (CSR) programs. In doing so, I argue that over the last two decades, neoliberal policies, public-private partnerships (PPPs), privileged access to bureaucratic knowledge, and the power to massage legal structures have restructured the permit process for stone quarries and gold mines. The ability of major companies to receive special privileges in host countries and to challenge (and change) legal structures and policies reflects the "corporate investment rights regime" enabled by "extractive exceptionalism."[18]

Scaling Up

Universal values and common global heritage often frame the rhetoric of international diplomacy agendas focused on heritage. In addition, initiatives by the United Nations support scaling up development agendas, including those of the extraction industry, arguing that they can improve social and economic mobility. On average, extraction projects are larger (at the time of this writing) than at any time in history.[19] From urban to rural landscapes, mega projects—mines, quarries, highways, bridges, tunnels, airports—offer new visions for what is now possible. Paul Ricoeur reminds us of Aristotle's perspectives on heritage and destruction: future growth is dependent on both destruction of heritage and construction of the new, the modern.[20] In the last two decades, heritage experts have applauded punctuated moments of technological change through inscriptions on the World Heritage list. The Cornwall and West Devon Mining Landscape (2006, United Kingdom) emerges from perfectly manicured grasslands; and inside the museum mine tools and spaces are exceptionally clean and polished. Gentrification projects have woven contemporary design and carefully presented aesthetic presentations of once harsh and dirty industrial landscapes at the serial property of Wallonia (2012, Belgium) and the Mines of Rammelsberg, Historic Town of Goslar and Upper Harz Water Management System (1992, Germany). In other cities, such as Lisbon (Tejo Power Station), enormous furnaces and turbines are backdrops for contemporary energy and science museums, even integrated into university libraries (Santral at Bilgi in Istanbul).

Lacking in all of these new "concept" spaces are the raw circumstances of the lived experience: the intense noise, steamy heat, and pungent sweat. The tough social and living conditions of the workers, many on the edge of poverty, are almost impossible to capture, even through historic photographs, correspondence, timesheets, and company plans. George Orwell's 1937 *The Road to Wigan Pier* (Penguin Books) does a better job, and parallels the current debates of scholars who argue that the politics of authority embedded in the recognition of industrial heritage signifies attempts to rectify unpleasant legacies of industrial growth that shaped global disparities of wealth. Here I turn to an overview of the Turkish context.

Turkish Mining

The power of international trade and mining concessions is not new to western Turkey. From at least the eighteenth century through the eve of World War I, access to mines and ports structured the interests of the British, French, Italians, and Greeks.[21] Concessions to resources were critical in negotiations during the 1919 Paris Peace Conference (see Chapter 2, this volume) and the mandates of the League of Nations.[22] Individuals and their state-sponsors sought to preserve private partnerships, largely owing to the prevalence of the co-mingling of familial, social, and economic ventures.[23] The business component of Greek concessions for mines in western Asia Minor was in part dictated by the quest for Greek sovereignty in the area.

Just as mining defined much of the negotiations after the fall of the Ottoman Empire, it also redefined how the Republic of Turkey structured relationships and private investments after 1923. The Public Expropriation Decree of 1925 gave authority for procedures to claim areas for civic improvement projects, including those determined by national cadastral initiatives in 1926. Over the next decade, efforts were made to integrate mines, roads, and railroads into regional planning. These developments made urban centers more and more dependent on rural resources.[24] At the Izmir International Fairs, as shown in Figure 1.3, extraction was promoted as a lifeline for Turkey's future economic capacity. By the mid-1930s, foreign concessions became harder to hold.[25] In 1935 the General Directorate of Mineral Research and Exploration was given jurisdiction to regulate many of the extraction industries. This followed in 1937 with legislation for the Turkish Iron and Steel Industries.[26]

Handsome US assistance packages (e.g., Truman Doctrine, Marshall Plan, Point Four, see Chapter 3, this volume) shifted the political and economic landscapes yet again in the late 1940s and 1950s. Foreign experts

FIGURE 1.3 "A view of a model of the Zonguldak Coal Mines in the Türkiye İş Bankası Coal Companies pavilion." 1936 Izmir International Fair. Courtesy of the Ahmet Piriştina City Archives, Izmir.

stepped into a new arena of consultancy. They advised on the development of the nation. Under Prime Minister Adnan Menderes' policies, the Turkish economy grew quickly. Even in the post-1960 coup environment and the wake of Menderes' death by military hanging, foreign assistance programs with the United States and Europe continued.

These connections resulted in Turkey's burgeoning ties as a potential future member of the European family. The European Economic Community (EEC) was established in 1958 (following the 1957 Treaty of Rome). Under the 1963 Ankara Agreement, Turkey moved toward a progressive Customs Union. Following the 1980 coup, changes in legal structures became more favorable to neoliberal arrangements (see Chapter 5, this volume), which followed those put forth by the EEC as well as the 1985 Integrated Mediterranean Programmes. Under the environmental legislation of the EEC passed in 1985 (Directive 85/337/EEC), Environmental Impact Assessments were required for members and those receiving support. Implementing legislation came in 1990 and further updates in 1997 (Directive 97/11/EC). In 1999 Turkey was invited to attend European Union (EU) meetings as a candidate for membership.[27]

Throughout this process, Turkish policies regarding trade became far more open, including regulations for mining. Turkish Mining Law No. 3213 was promulgated on June 15, 1985 (Official Gazette No. 18785) and paved the way for more fluid domestic and foreign capital investment with special provisions to increase opportunities for multinational corporations and transnational corporations. The sector was privatized in 1994. While ownership of subterranean resources remains vested with the state, long-term concessions give the private sector tremendous power.[28] Turkey is not the only country to have followed this trajectory.

International Mining

Analyses of the extraction industry in Ecuador, Mongolia, Nigeria, the Philippines, South Africa, and Tanzania shed light on profitable corporate structures.[29] Local infrastructure and communities are often given a back seat, contrary to the stated intensions of World Bank agendas to foster small-scale mining and open dialogue.[30] In fact, the Millennium Development Goals and the European Commission have promoted scaling up mining operations in Turkey through PPPs.[31] Through this practice, communities and lands caught between territory and extraction regimes become, following Lindsay Shade's scholarship, "sacrifice zones" between "the sovereign and the subsoil."[32] The prestige of the foreign counterpart leverages capacity-building narratives that purport to support the pursuit of social mobility. How communities and individuals choose to live is no longer a personal choice: it is entrenched in rural capitalism and profit-driven metrics.[33]

International frameworks claim to deploy measures to avoid such situations. In 2001 the International Monetary Fund (IMF) initiated the Reform Programme for Improvement of Investment Environment in Turkey. The result was the 2004 IMF-Mining Law, which sought to accelerate mineral exploration by the private and foreign initiatives by eliminating thick bureaucracy; in so doing, countries have prioritized their economic sectors and given less attention to environmental protection and social life, contrary to recommendations by the UN Committee on Economic, Social, and Cultural Rights and the UN Committee on the Elimination of Racial Discrimination.[34] The UN Global Compact, a group of CEOs, also aims to promote progressive CSR programs under the Sustainable Development Goals[35] Canadian companies' CSR performance standards have become the baseline for the European Commission and the Extractive Industries Transparency Initiative. As Barry has shown for Democratic Republic of Georgia and pipeline contracts, however, transparency can be a very loose concept.[36] Companies quickly adapt, (re)invent, and evolve their agendas to fit current trends and regulations, [37] and in so doing, find new ways to secure their unofficial, yet imperative, social operating licenses.

In western Turkey, management of forests and the registration of archaeological areas are unstable nexus points where we can trace how perceptions of power are often sufficient to ignore best practices and current legislation, and how powerful constituencies can overturn legal parameters. It is for this reason that I now review regulations for the zoning of forests and archaeological sites as well as the practice of eminent domain in Turkey.

Productive Forests

Under the Tanzimat reforms (e.g., the Monetary Reform Act of 1844), foreign investors (e.g., British, French, Italian, Greek, and American) were allowed to pursue concessions for mining and related infrastructure (e.g., railroads and ports). In some cases, they aggressively revived Ottoman mines for their own purposes.[38] Further regulations were put in place in 1870, after the establishment of the General Directorate of Forestry.[39] In the following decades, a close relationship emerged between the management of forests and mining, overseen by the Ministry of Finance[40] (see Table 1.1).

Beginning in the 1920s, legal instruments pertaining to forests, natural resources, mapping, and expropriation were organized under the new Republic.[41] Legislation was revised in 1937 (Forest Law No. 3116) and again in 1945 (Amendment Law No. 4785). Changes in 1950 streamlined mapping activities in rural areas, and Land Registry Law No. 766 sought to harmonize regulations for expropriation. Again, jurisdiction was revised under the 1956 Forest Law No. 6831, which outlined greater specificity for the boundaries, development, and use of forests. Formal management of forests was proposed in 1963, corresponding to the establishment of the Ministry for Energy and Natural Resources, and, in 1969, the Ministry of Forestry.

More revisions took place in the 1980s, including the 1983 Expropriation Law No. 2942, as amended by Law No. 4650.[42] Harmonization for rural development moved further along in 1987 under Cadastral Law No. 3402. Importantly, however, authority for forest areas was vested directly with the General Directorate of Forestry and included provisions for areas abutting forest regions, giving this General Directorate immense power.[43] In 1990 Turkey established the Ministry of Environment, which was merged with the Ministry of Forestry in 2003. Less than a decade later, in

TABLE 1.1 Key Ministries (and Protection Councils) Related to Management of Forests and Cultural Heritage in Turkey

MINISTRY NAME	DATE
Ministry of Education (Antiquities and Museum Directorate)	1933
Ministry for Energy and Natural Resources	1963
Ministry of Forestry	1969
Ministry of Culture	1971, 1974–1977
Ministry of Education and Culture	1977
Ministry of Culture	1977–1982
Regional Protection Councils established	1987
Ministry of Culture and Tourism	1982–1989
Ministry of Culture	1989–2003
Ministry of Environment	1990
Ministry of Environment and Forestry	2003
Ministry of Culture and Tourism	2003 (to present)
Ministry of Forestry and Water	2011
Ministry of Environment and Urban Planning	2011

2011, a reshuffle resulted in the formation of the Ministry of Environment and Urban Planning and the Ministry of Forestry and Water. Some argue that recent legal structures have favored development (mining, quarrying, planting, etc.) over human rights. These scholars point to cases of compulsory expropriation, which have been contested in the European Court of Human Rights.[44]

Today the Ministry of Forestry and Water Affairs oversees the General Directorate of Nature Conservation and National Parks, General Directorate for Water Management, and General Directorate of Forestry (among other units). Since 2008 management plans have attempted to regulate more seriously logging, hunting, tourism, and extraction in national parks and forest areas.[45] Under Forest Law No. 6831.4999 (5/11/2003), the General Directorate of Forestry grants long-term leases for resource extraction, which they gamble will result in pleasing profits and thus favorable returns to the state. This model embraces investment.

The expansion of extraction in forests results in shifts to not only how forests are managed at the ministerial level, but also how forest legislation complements regulations for extraction, notably recent amendments to Mining Law 2005 (Law No. 3213), 2010 (Law No. 5995), and 2015 (Law No. 6592). These regulations streamline applications for prospecting and formal concessions in forests, offer incentives for multinational partnerships, simplify environmental impact studies, and make possible compulsory expropriation of private land.[46]

Caring for Culture

Given the increased pace and scale of extraction in forests, and the revisions to mining and forestry legislation, cultural property laws and the registration of heritage sites are more pertinent to this discussion than ever. The first Ministry of Culture in Turkey was established in 1971; prior to this point archaeological heritage management was situated in the Ministry of Education. In 1973, Law 1710 (*Eski Eserler Kanunu*) introduced a firmer definition of protected zones (*sit alanı*). This legislation focused not only on monuments but also on surrounding and adjacent areas. In addition, guidelines were submitted to the High Council of Real Estate Antiquities and Monuments (*Gayrmenkul Eski Eserler ve Anıtlar Yüksek Kurulu*, GEEAYK) to consider protection of historic sites and zones. In 1983 legislation shifted again under Law No. 2863 for the Conservation of Cultural and Natural Properties. In 1987 the Regional Councils for Conservation of

Cultural and Natural Properties (*Yerel Koruma Kurulları*) were introduced; each regional council oversees site investigation and registration and coordinates with the Directorate of Land Registry (*Tapu Müdürlüğü*). Councils conduct their work collaboratively with regional museums and report their decisions to the General Directorate of Cultural Assets and Museums in Ankara.

Between 1977 and 2003 the tie between culture and tourism under a single ministry waxed and waned, but since 2003 these two ministries have remained linked. In 2004, Article 7 of Law 5226/1 introduced the concept of a "management site" and called for coordination of natural and cultural landscapes with urban and rural planning offices. Also under this legislation, archaeological areas could be designated as multilayered landscapes. However, even with this expanded definition, the ability to balance access to public resources with transference to private, highly valued financial resources remained unclear. Archaeological sites are public goods with little opportunity for lucrative leverage stemming from research agendas alone, and so far Turkish legislation forbids the sale of excavated material in or outside of Turkey. In contrast, while natural resources (e.g., marble, granite, gold, and copper) are public goods, concessions for their extraction have a favorable, free, and immediate market component.

Narratives of Extraction

With greater context of the stone and gold industries at play in Turkey, the case studies introduced previously on marble and gold deserve further treatment. In order to broaden the understanding of the Pomza gold mine in Sart, I delve into the corporate success and professional standards of Tüprag's mega-mine in nearby Uşak, particularly the important role of their CSR programs. In contrast, the thwarting of a marble quarry by the community of Hacıveliler was largely a result of the community's conviction to block the initiative, evidence of archaeological remains, the prestige of international tags such as World Heritage and the IUCN, and the marble company's lack of CSR.

A Mega-Mine

Uşak is famous in archaeological circles for the sixth- and fifth-century-BCE gold and silver "Lydian Treasure" that was repatriated to Turkey in 1993 after having been spirited out of the country in the 1960s. Most of the exquisite objects were plundered from Lydian tumuli, or burial grounds, in the area. The material resurfaced in the 1980s as part of the New York Metropolitan Museum of Art's exhibition, *A Greek and Roman Treasury*, confirming that the Met had purchased many plundered items in the 1960s.[47] After litigation, the material was returned to Turkey in 1993 and later transferred to the Uşak Museum. The well-illustrated publication *The Lydian Treasure: Heritage Recovered* reviews the case and presents a catalogue of objects made of silver, gold, and precious stones. By extension these objects speak to the practice of locating, extracting, and refining natural resources in antiquity. Despite the fame of Lydian power derived largely from its wealth in gold and silver,[48] contemporary prospecting in Uşak did not begin in earnest until relatively recently.

In the late 1980s, Eurogold, a joint venture between Australian and Canadian firms, began prospecting for gold in Bergama. Legal action against the endeavor was successful in the courts, yet power and networks made enforcement impossible. The operation went forward. Subsequent complaints, news articles, and lawsuits claiming environmental degradation and community disapproval only emboldened the industry, ultimately leading to legal changes that favored corporate expansion.[49] Other companies followed suit.

Founded in Ankara in 1986 (i.e., a year after the 1985 Mining Law 3213, see previous mentions in this chapter), Tüprag Metal Madencilik Sanayi ve Ticaret Anonim Şirketi is a 100-percent subsidiary of Eldorado Gold Corporation in Vancouver, Canada. The company operates the Efemçukuru (Izmir) and Kışladağ (Uşak) gold mines. They delight in the history of Lydian gold: the company's logo is the head of a lion, copied from Lydian coins appearing as early as the seventh century BCE, as shown in Figure 1.4. In their recent brochure, "Sustainable Gold Mining in Europe," the European Association of Mining Industries, Metal Ores & Industrial Minerals celebrates the ancient Lydians for their development of "the yellow metal."

The Kışladağ mine in Uşak is not well known by the general public in Turkey, despite being the largest gold mine in Europe and having a capital investment of 350 million USD. Prospecting in the region began in the late 1990s. Environmental assessments, expropriation of private lands, and a lease on land designated as forest followed. Operations for commercial production began in 2006. Over the last decade, this open cast mega-mine, approximately 850 hectares in area (roughly 8 square kilometers, or 2,100 acres), has displaced thousands of cubic meters of earth and in the process created enormous earthworks that descend into the ground, known as the

FIGURE 1.4 Gold Stater with Lydian Lion. Courtesy of the Museum of Fine Arts, Boston. Accession number 04.1541.

"the pit,"[50] as well as rise from it. In fact, the colossal "clean" earthworks (i.e., crushed stone and soil without gold) will soon be taller than the surrounding natural peaks, a point of great pride for workers at the mine.

Extracted rock with embedded gold is crushed and deposited on the now monumental "leach heap." This 2.5 kilometer by 1 kilometer (and hundreds of meters tall) rectangular man-made pyramid of earth sits on an impermeable lining. A system of pipes facilitates the leaching process whereby a cyanide-lime mixture is drizzled over "the heap." The lime neutralizes the mixture enough to make the air breathable, while the cyanide leaches the gold from the rock, conveying it to the collecting tanks.[51] The toxicity level is significant: during my tour, I was told that drinking one liter wouldn't kill a person, but consuming 20 liters certainly would. The public relations officer said that the leach heap "must never be touched, must never be moved." A recent study claims that it will take 585 years after closure to complete full mitigation measures.[52]

Despite the colossal amount of toxic waste it produces and contains, Kışladağ has been able to move away from negative imagery associated with gold mines. This is in large part due to their substantial CSR programs. Approximately 80 percent of the local workforce resides in the province of Uşak. The company's brochure states that "seasonal subsistence farmers" are now "partners of the mine." Just as "redoubling" economic models pushed mid-twentieth-century water diplomacy in the

Gediz Valley (see Chapter 3, this volume), this resource has galvanized immediate, local change. In the eyes of municipal and government offices, Tüprag's promised "economic multiplier effect" gives long-term benefits through transferrable skills and knowledge, for employment either at other mines or in parallel sectors. According to various reports and recounted during tours at the mine, Tüprag has patronized local schools, redone local water systems, paved roads, and even supported the new archaeological museum in Uşak, where the Lydian treasure will be a highlight.

According to Özen and Özen, the CSR programs for Kışladağ were launched prior to opening the mine, and thereby propagated a working relationship between Tüprag and the local community well in advance of major operations. On the one hand, this type of structure perpetuates socio-economic difference and engenders a top-down corporate culture; on the other hand, the transformative change in economic mobility creates local buy-in, despite resistance during contentious negotiations for land. Court orders were issued to close the mine in October 2007, but, under claims of national interest, Ankara moved forward with compulsory expropriation of privately held lands and a long-term lease on forestland.[53] Tüprag officials claim that to help forge positive relationships with the community, higher-than-market prices were paid for land.

A number of important points should be considered with regard to the Turkish gold sector. Multinational corporations value their networks with top-ranking bureaucrats and politicians, yet efforts toward and investments in local development through CSR have proven to be extremely important for mitigating negative press and managing local discontent. Additionally, the momentum generated at Kışladağ has implicitly provided support for the initiatives of Pomza Export at Sart, Turkey's smallest and oldest gold operation, described to me by those at Kışladağ as Turkey's "boutique mine."

A Boutique Mine

Brochures from Kışladağ were shared with a self-made and self-proclaimed highly successful organic farmer in the district of Salihli, Manisa. She was unfamiliar with the mine. She studied the brochures and, with grave concern, asked if the operations at Pomza used the cyanide-leaching technology and if Pomza could have a negative impact on her children, on her community. She was told the gold is placer gold, and thus its extraction requires a far less toxic process. There isn't the potential threat of colossal contamination that there is at a facility with a heap-leaching system. The farmer welcomed this news. She then asked how close it was to Sardis.

The municipality of Salihli appropriates the historical cache of Lydian gold and Sardis for its own cultural capital. Just outside of the bustling zone of the largest outdoor market stands a larger-than-life statute of a gold worker. From a long ladle-like tool he pours molten gold that miraculously turns into oversize Lydian coins once they hit the platform that surrounds his feet. Frequently stolen from the monument, these faux coins have taken on value of their own, perhaps related to the civic awards given by the Salihli Chamber of Commerce and the Municipality, which also feature large, faux Lydian gold coins embossed with the Lydian Lion presented in red velvet boxes. Local historians and municipal offices have also embraced the performance of scholarship, hosting the 2015 October gold symposium at the luxury Lidya Sardes spa hotel.

The hotel capitalized on the iconic bath gymnasium motif at Sardis, which has also been incorporated into stamps, appropriated for restaurants and shops, used as screensavers, and featured on municipal publications and strategic plans, including the main website for MARBLE 22. Moreover, the hotel adapted the Sardis concept. Conference rooms and areas in the hotel carry the names of monuments at Sardis and related cultural landscapes: the Sardis meeting room, the Tmolus and Kybele restaurants, the Pactolus Pool, the Pactolus Bar, and the Dionissos Disco. Throughout the hotel, design concepts for mosaics, murals, paintings, and sculptures are drawn from those of the excavations. In fact, the hotel had an in-house workshop for residence artists to work directly from the expedition's publications. For these reasons, many believed that the legacy of Sardis would surely thwart Pomza.

According to reports and interviews, the decision to proceed at Pomza was made in consultation with the Izmir Second Protection Council, the Manisa Museum of Ethnography and Archaeology, and the General Directorate of Mining Affairs. Despite no consensus among this expert group, the interests of the General Directorate of Mining Affairs prevailed. A critical point came in the year 2000. The exact nature of zoning for the Necropolis (i.e., as a historic monument) wasn't clear. That the Necropolis was part of a larger ensemble of the Sardis multi-layered historical landscape didn't matter; at the time (i.e., prior to the 2004 revision to the law) areas were considered individually (e.g., the Temple of Artemis and Church, the Bath Gymnasium/Marble Court/Synagogue, the gold refinery, Roman houses, the Lydian wall, and the Acropolis).[54]

Responding to the case, the Archaeological Exploration of Sardis sought to thwart future operations on the grounds that the entire area comprised archaeological heritage of significant value to Turkey. Protecting this heritage, however, required payment to Pomza, and litigation against the expanding and powerful gold lobby. The company had invested capital in basic infrastructure as well as time.[55] To walk away from the operations (in 2002) required major compensation (upward of a quarter of a million USD). Sufficient funds were not available on such short notice. Others lamented that the entire undertaking violated the principal of the national patrimony law and the ideals of preservation. Pomza subsequently secured a lease from the Treasury and the Ministry of Forestry.

Sixteen years since the first backhoe sighting in the fall of 2000, as described in the introduction of this chapter, the community has accepted the mine, with many locals exhibiting pleasure with the full-time employment it offers and the contributions to local infrastructure through CSR, such as support to local schools. Contamination to the environment has been minimal because the process for extracting placer gold requires hydraulic techniques that, while they may produce an abundance of clayey silt, do not require cyanide or other toxic materials. Tourists have continued to visit the region, often staying at the Lidya Sardes Hotel, and municipal publications have continued to celebrate their favorite muses, the Bath Gymnasium and Lydian Gold. For local communities, Pomza has not been a problem, but rather an asset.

The transformation to the historic landscape, however, has been substantial. On a clear day from the tumulus of Alyattes in Bin Tepe—the royal burial grounds of Sardis some 15 kilometers (9.5 miles) north across the Gediz plain—the gash in the Necropolis spur is apparent. In my conversations and interviews with Pomza's representatives as well as residents in the region, they've shown little concern about the shifting appearance of the region's topography. Yet, everyone has been resolute that the history of gold, and therefore (in their minds) Sardis, belongs to the district of Salihli and that this legacy should be vested with the community of Sart and the Archaeological Exploration of Sardis. They've expressed great interest in collaboration with the Archaeological Exploration of Sardis, a museum in Salihli, and the rehabilitation of the ancient gold refinery and landscaping around the Pactolus River.

The ongoing activities of Pomza and the ongoing destruction to the integrity of the historic landscape, both physically and symbolically, came up during the process of preparing the file for Sardis' inclusion on the Turkish tentative list for World Heritage sites. Although Turkish legislation does not include legal instruments for buffer zones of heritage areas, sites on the World Heritage list must conform to these standards. Representatives from the cultural sector were not interested in broaching the topic precisely because it wasn't necessary for the tentative list—yet nor was the tentative listing reason to acknowledge and to begin to deal with the white elephant in the room. Should a dossier for full inscription of Sardis (and Bin Tepe) be submitted, the operations of Pomza will need to be explained.

Compared to Kişladağ, Pomza is tiny, but in the context of the village of Sart and the archaeological site of Sardis, the operations of Pomza are immense, and have moved quickly. The earthworks and infrastructure have become features in their own right. They are indicative of transnational connections so powerful as to potentially render cultural heritage obsolete. In 2014 Pomza made a request to expand operations. The Second Protection Council in Izmir revisited the territory of the Necropolis and determined that the mine was encroaching on archaeological remains and could further destabilize the area. The very top of the Necropolis was to be rezoned with the highest level archaeological protections possible.

Two years later, the archaeological zoning for the Pomza and Sardis territories was revisited again. On December 8, 2016, Decision No. 45.07/16 was released to the public.[56] Archaeological zones had been revised and boundaries updated.[57] The argument centered on Article 7 of the National Patrimony Law (2863, as amended on July 14, 2004, No. 5226.1)—recall the discussion of this law earlier in this chapter. As envisioned, the law seeks a more holistic definition of an archaeological zone, incorporating the entire ensemble of monuments and associated landscapes; it also envisions integrative conservation and management plans. As with all Protection Council decisions, this one was sent to the various municipal offices in Manisa, as well as official offices pertaining to water, culture and tourism, natural resources (including mines), forestry, and roads—and it was posted online for the general public. According to the new decision, all future activities within the newly defined zones would require permission from the Directors of the Archaeological Expedition of Sardis and the Manisa Museum of Ethnography and Archaeology. The area extended west to the closest town (Mersindere) and east to the other side of the Acropolis (the Bozdağ access road) and included the entire Acropolis, its primary monuments, the village of upper Sart Mustafa, and the Necropolis. But mining didn't stop, nor were there plans for immediate closure. In February 2018, the ruling was reserved.

Just as homes within the protected archaeological zones of Sart are grandfathered, so, too, was the infrastructure of the mine—the settling tanks, the offices, the roads, and the extraction zones. The biggest loophole is found along the southern slopes of the Necropolis, toward the Tmolus mountains. There is one point where the buffer zone does not extend beyond the highest level (first degree) archaeological zone, thus giving access to the southern slopes of the Necropolis and the foothills of Tmolus beyond. Therefore, while this ruling acknowledges the archaeological significance of the area, it also provides space for business as usual, especially for gravel. The mine has the contract for the gravel bed of the high-speed train between Izmir and Ankara. The irony is very rich: for the first rail-line bed in the nineteenth century, marble seats from the Hellenistic theater were crushed. This type of secondary mining for gravel is also common among limestone and marble quarries. Interestingly, the timing of these negotiations (2014–2016) overlapped with the case in Hacıveliler, near Kaymakçı, and the proposed marble quarry, approximately 20 kilometers (12 miles) northwest of Sardis, where I began this chapter and to which I now return.

Grazing in Quarries

This chapter started with the case of a proposed marble quarry in public (*mera*) and forest (*orman*) land used by the community of Hacıveliler for grazing. Situated between the current village along the lakeshore and their ancestral home in the mountains, this land is of symbolic and physical importance to the community. As explained previously, lawyers leveraged archaeological zoning after arguments of traditional landscapes and lifestyles failed in the courts. In contrast to the Salihli and Sart communities, the Hacıveliler community had shown no substantial interest in the archaeological heritage of their immediate landscapes. They were also indifferent to annual work offered by the Kaymakçı Archaeological Project, preferring farming and pastoral activities. What is more, their approach to managing access to their ancestral village had been through calculated ruination.[58] Community members had also been indifferent to archaeological zoning for the area of Kaymakçı (begun in 2011), Bin Tepe (and its 2013 World Heritage Tentative listing), and the 2009 IUCN designation of the Important Bird and Biodiversity Area for Lake Marmara. Yet, they were not indifferent to increased activity at the would-be quarry areas in March 2015, and they were furious that their claims to traditional lifestyles were unrecognized by the courts.

Stone companies in Turkey meet the basic regulations to manage and contain debris, but they frequently neglect a comprehensive perspective on rural aesthetics and lifestyles. Even so, it is nature and culture that most often frame their approaches to branding. Slogans such as "from nature to your building," "natural lifestyle," and "a gift from nature" illustrate the stark contradictions, as do logos of stylized mountains and variations on the design of ionic columns. Historic name recognition (e.g., Assos, Ayasofa, and Palmyra) and the historic value of marble from quarries used in famous monuments, such as the Library of Celsus at Ephesus, is also a significant component of the industry's advertising campaigns.

Despite these public displays of historical connections, natural stone companies provide almost no support for archaeological research, museums, or other related fields. Interviews with twenty-two companies at MARBLE 23 revealed a lack of interest in corporate philanthropy, regardless of its mission (i.e., arts, sports, or education). A review of sixty-four brochures distributed in exhibit halls, followed by subsequent review of those companies' websites, found only one stone company with a philanthropic mission. Pet Holding supports contemporary artists (painters and sculptors), art exhibitions, and the performing arts (concerts and theater). This company began in the petroleum business in Iraq, Azerbaijan, Kazakhstan, Yemen, and Turkey. Only in 2010 did they move into Turkish marble. Unlike nickel, copper, gold, or petroleum industries, quarrying does not require (nor does it produce) inherently toxic materials. As Fiona Starr indicates in her research on CSR and heritage in South East Asia,[59] industries perceived to be more harmful to the environment than others often have vigorous programs to garner greater respect with local communities.

In this global context of extraction, when a dozen or so women from a very small village set out to thwart the proposed marble quarry, they chose to take on a major industry. The marble company seeking the concession for marble near Hacıveliler had made no effort to arrange for a community meeting. They had offered no viable social programs. Rather, justification for the concession was purely profit-driven for the company. The small agro-pastoral community grazing sheep and goats contributed to the regional economy at a scale too insignificant to justify blocking the quarry. Undeterred by the court's decision, the legal team pushed forward with greater emphasis on matters of preservation, explicitly pressing the international benchmarks of World Heritage (buffer zones for Bin Tepe and by extension Sardis) and the IUCN (the IBA listing of Lake Marmara).[60] Subsequent to the July 2015 court ruling in favor of the archaeological zoning, on April 29, 2016, the decision was issued to extend the first-degree archaeological zone, thereby preserving the local landscape and preventing new quarrying activities.[61]

While it is tempting to see local resistance and heritage badges as the primary platforms to realizing success in this case, the lack of a CSR program and of efforts to engage with the community were also contributing factors. Stone companies may draw on antiquity and contemporary heritage for branding purposes, but they rarely show a sustained ethnical or intellectual interest in heritage. Economic opportunism is the primary motivating factor, and barring requirements or commitments to CSR programs, we find an industry carving up the mountainsides of Turkey to please foreign markets. The irony is that their elite clientele in the United States and China are also among the primary visitors to places such as Ephesus and Sardis and their marble facades—the same contingent highly dependent on global financial security made possible by the gold standard.

Discussion

This chapter demonstrates that state strategies for access to subsoil resources create a suite of tensions. In the province of Manisa in western Turkey, the extraction industry relies on access to territory through concessions for forests predicated primarily on economics and short-term return on investment. Family farms were benchmarks of success for development diplomacy in the mid-twentieth century, but today farmers and herders have been reclassified as inexperienced laborers. Mega-mines, in contrast, offer transferable skills applicable in the mining, quarrying, and construction industries. Steady contracts mean reliable paychecks and robust benefit packages, which in turn result in relatively higher standards of living than what is currently possible for small family farmers.

These processes have another consequence: the rise of industrial farming spurred by neoliberal and rural capitalism. As the pauperization of the family farm becomes a reality, more and more people are driven to work for mines. This offers opportunity to expand farms considerably, and the EEC would like to see this level approach the EU corporate (family-run) farm model. As researchers have shown, CSR programs today focus on strategic long-term changes. Companies are seen as catalysts for corporate and social change in rural areas—Roy's "bottom billion capitalism" and "neoliberal populism."[62] Social scientists warn of social and economic marginalization driven by state authority and global corporatism. In these ways, the entanglements of preservation politics and transformative landscape modifications are both incredibly disruptive and exceedingly connective.

These case studies provide a window into how communities in the wider Izmir region, bringing in the provinces of Manisa and Uşak, react to extraction, and how archaeology becomes entangled in these debates. The legislative changes to archaeological areas in 2004 under Article 7 pledged holistic management. The July 2015 ruling in favor of archaeology at Hacıveliler and the surrounding area, including historic quarries, as well as the 2016 updated zoning areas for not only Kaymakçı but also Sardis and its Necropolis, indicated major changes to how archaeological territories were to be managed. The reserval for the Necropolis (February 2018) demonstrates the fluidity of these types of cases. Time will tell if these regulations are strong enough to keep the natural stone and gold industries at bay. Yet, as I discuss in the following chapters, the extraction industry is only the latest economic pursuit in a long history of strategic business opportunities that involve foreign counterparts operating on Turkish soil in Izmir and its hinterland of the Gediz Valley. In the next chapter, I turn to the thick web of late-nineteenth-century and early-twentieth-century foreign economic interests in Turkey that came with both benefits and failures for how diplomatic endeavors and economic opportunity confronted heritage landscapes.

CHAPTER 2 | Preference in Paris

Before these negotiations begin the United States should determine what solution of the Turkish problem it desires, so that our representatives in the making of the Turkish Treaty may know from the very start the general outlines of our policy. If not provided with such a programme, they will be at great disadvantage in dealing with the British, French and Italians, whose aims are definite and are sure to be elaborately worked out.

—MEMORANDUM OF THE POLICY OF THE UNITED STATES RELATIVE TO THE TREATY WITH TURKEY. Confidential document submitted by William H. Buckler on behalf of the American Commission to Negotiate Peace. Hotel Crillon, Paris, November 26, 1919.

There is a marked spirit of depression; last hopes are centered on the London Conference. Halil Bey asked if the Allies are bent upon the utter destruction of Turkey, for only thus can he interpret their attitude. It appears to be the common opinion that Turkey's chances for the future rest upon the ultimate disposition of Smyrna. With Smyrna a peaceful and national development of Turkey-in-Asia by the Turks and for the Turks is possible: without Smyrna, fruitless dissention and eventual disintegration are thought inevitable

—WILLIAM R. BERRY, Sardis Report, February 26, 1921

CHAPTER I DEMONSTRATED THE TRANSFORMATIVE twenty-first-century extraction agendas that have engulfed the hinterland of Izmir and threatened archaeological sites. This chapter also explores development, foreign investment, and archaeology, but from the perspective of the late -nineteenth and early twentieth centuries. This was Smyrna's heyday. She offered direct access to foreign lands through her port and, through well-trodden

FIGURE 2.1 H. C. Butler in excavation tent. No. C.7.5.d. American Society for the Excavation of Sardis 1910–1914. Courtesy of the Department of Art and Archaeology, Princeton University.

routes and new railroads, ease of access into the heartland of Anatolia and beyond. These opportunities propelled an entrepreneurial spirit in the city.[1] At the turn of the twentieth century, American investors were among the most persistent, seeking to edge out the surreptitious pacts of the British, Greeks, French, and Italians.[2] American archaeologists set their sights on Sardis.

Between 1910 and 1914 Sardis became the muse of a group of Princeton men who shared a love for antiquity and a devotion to philhellenism. Their efforts in the US and western Anatolia coalesced with the Princeton Expedition to Sardis and the establishment of *The American Society for the Excavation of Sardis*. Howard Crosby Butler was a faculty member at Princeton and director of excavations.[3] William Hepburn Buckler was a diplomat, lawyer, and archaeologist (his love was epigraphy). In 1910 Butler commanded operations at Sardis from a very nice excavation tent, as shown in Figure 2.1. In short order, Villa Omphela, shown in Figures 2.2 and 2.3, was built to serve the expedition: residence, eating, library, museum, and depot. It was humble for Butler, but for locals, as Butler wrote to colleagues and family, it was considered a palace, complete with

FIGURE 2.2 Butler's Excavation House, Villa Omphale. No. C.7.7.f. American Society for the Excavation of Sardis 1910–1914. Courtesy of the Department of Art and Archaeology, Princeton University.

FIGURE 2.3 Inside Village Omphale. No. SAR_AL_004. American Society for the Excavation of Sardis 1910–1914. Courtesy of the Department of Art and Archaeology, Princeton University.

FIGURE 2.4 Tennis group on the Sardis court. No. SAR_AL_001. American Society for the Excavation of Sardis 1910–1914. Courtesy of the Department of Art and Archaeology, Princeton University.

a tennis court, as illustrated in Figure 2.4. In their entirety, then, the research, residential, and recreational facilities built on-site positioned Princeton's Sardis expedition as a colonial enterprise on foreign soil in pursuit of ancient civilization.

Excavations paused at Sardis in 1914 due to World War I. Owing to the uptick in tensions between the Allies and the Ottoman authorities regarding Asia Minor, especially the city and countryside of Smyrna, Butler and Buckler developed contingency plans to secure Villa Omphale and the Temple of Artemis, which had been their primary focus. They also pushed the US Archaeological Institute of America (AIA) to pursue an international body that would oversee archaeological sites and antiquities in the eastern Mediterranean and the Near East. From 1917 to 1919 Sardis itself took on great symbolic value and geographic importance. The Allies supported the Greek occupation of Smyrna and considerable areas of her hinterland. A barbed-wire fence along the Sardis Acropolis marked the Greek-Turkish frontier, and thus the site became *the* eastern boundary of Greek occupation.

Philhellenism, the perception of "democracy," and, especially, economic promise defined the political posturing of the Allies. The "Big Three" (David Lloyd George, George Clemenceau and Woodrow Wilson) steered the agenda for the 1919 Paris Peace Conference. Over the year-long deliberations the delegation attempted to grapple with and reconcile their impending post-colonial world. The goal was to draft and promulgate the Treaty of Sèvres, including the tenets of the League of Nations. This process advanced conceptions of national identity and, in so doing, also demanded new models in order to achieve the delegation's desired spheres of economic expansion. Their proposals represented hybrid approaches that spun future agendas in their favor, thereby re-inviting as well as perpetuating modes of imperial violence.

It was during the negotiations in Paris that Buckler lobbied for the international commission on antiquities. Stateside, Butler rallied support with the AIA and other US institutions as well as learned societies in the United States; Buckler pursued these relationships in Britain and France. These efforts complemented those of Edward Capps at the American School of Classical Studies in Athens (ASCSA), who jockeyed carefully with Greek government officials as well as US diplomats (i.e., George Horton, see later in this chapter) to extend the arc of ASCSA into Asia Minor.[4] Together these various groups attempted to sway US and European policy toward a specific vision of archaeological research and site management. In these ways colonial networks of power and prestige framed the Western gaze such that the resulting sentiment reinforced feelings of entitlement that underwrote a quieted last gasp of claims to the heritage of Asia Minor in the wake of World War I.

Pedigree and family were vital in this endeavor. Buckler's half-brother opened many doors. Henry White was a career diplomat who had served as US Ambassador to Italy (1905–1906) and France (1906–1909). He was appointed the American plenipotentiary to the 1919 Paris Peace Conference.[5] It was through White's contacts and Buckler's society networks and knowledge of Asia Minor that the issue of archaeology in Anatolia gained traction in Paris. When White became ill (August and September of 1919) the American delegation to Paris relied even further on Buckler's perspective regarding not only antiquities, but also the wider "Turkish problem." During this process, as gleaned from Buckler's personal correspondences to his wife, Georgina Grenfell Walrond, and their daughter Lucy, he hobnobbed with Europe's cultured elite, including Arthur James Balfour, Henry Morgenthau, Hugh Campbell Wallace, Edith Wharton, A. I. Whyte; royalty, such as Russian Prince and Princess Kotschoubey; and key businessmen, such as Mr. Cyrus H. McCormick, Jr., whose company International Harvester had a growing footprint in Smyrna. First over informal lunches, dinners, teas, and drives through the countryside, and later in conference sessions, Buckler also met with prominent heads of state: French Prime Minister Georges Clemenceau, British Prime Minister Lloyd George, and Greek Prime Minister Eleftherios Venizelos.

Arnold Toynbee, a member of the British Political Intelligence Department and in charge of Ottoman Affairs in Constantinople, also offered critical perspectives. Toynbee's experience included a brief period at the British School at Athens prior to his work during World War I with the Political Intelligence Department of the British Foreign Office. He would later be posted (1921–1922) as correspondent for the Turkish War of Independence by the *Manchester Guardian* and remembered for his 1922 publication, *The Western Question in Greece and Turkey: A Study in the Contact of Civilizations*.[6] The "Eastern Question" and the "Turkish problem" were only questions and problems, in Toynbee's analysis, for those who wanted to control specific areas and resources in these territories—those who had interests in finance (i.e., banking), infrastructure (e.g., railways), agriculture (i.e., "mechanical farming"), extraction (e.g., copper, mercury and coal), and classical archaeology. His analysis bears truth for our understanding of the Smyrna-Sardis relationship in the context of the 1919 Paris Peace process and the establishment of the League of Nations. Archaeologists, collectors, and diplomats pursued specific policies for not only Smyrna but also her hinterland, where Sardis was located.

The Paris Peace Conference, League of Nations, and the Treaty of Sèvres

Susan Pedersen's reassessment of the League of Nations demonstrates how the Great Powers re-envisioned old models through new narratives for personal gain.[7] Patricia Clavin agrees, arguing that the League's primary focus was not to promote international peace, but rather to facilitate favorable economic investment policies. In essence the League provided a smokescreen to structure advantageous arrangements for the Allies.[8] The diplomatic dances at play during the 1919 Paris Peace Conference, then, were aimed at strategic modes of conference diplomacy, whereby key state actors sought to influence politics and trade deals, especially through future mandates.

Initially it was the Treaty of Sèvres that set boundaries for the partitioning of the Ottoman territories and the tenets of the League of Nations. The drafters envisioned the League as a collaborative international body with a primary goal of mitigating future conflict:

> In order to promote international co-operation and to achieve international peace and security
>
> by the acceptance of obligations not to resort to war,
>
> by the prescription of open, just and honourable relations between nations,
>
> by the firm establishment of the understandings of international law as the actual rule of conduct among Governments, and
>
> by the maintenance of justice and a scrupulous respect for all treaty obligations in the dealings of organised peoples with one another. . . .

These principles built on US President Wilson's speech of January 8, 1918, where he outlined fourteen points for world peace and a vision for the League of Nations. Further analysis, however, reveals that much of the rhetoric focused on development diplomacy and assistance programs would ultimately line the pockets of Western countries.[9]

The League legitimized this approach through the structure of Mandates. Article 22 of Sèvres defined the terms:

> To those colonies and territories which as a consequence of the late war have ceased to be under the sovereignty of the States which formerly governed them and which are inhabited by peoples not yet able to stand by themselves under the strenuous conditions of the modern world, there should be applied the principle that the well-being and development of such peoples form a sacred trust of civilisation and that securities for the performance of this trust should be embodied in this Covenant.

As the diplomats and politicians in Paris navigated the details of the Sèvres treaty, others questioned their intentions. Too much encroachment on sovereign territory would result, they feared, in a humanitarian crisis, yet without an advisory body, hostile actions toward minority groups and their property would run rampant.[10] At first President Wilson supported the sovereignty of Turkish areas of the Ottoman Empire, save Constantinople.[11] Debate arose over a US Mandate for Smyrna and her hinterland. In opposition to the voice of Turkish partners, Allies shifted their positions and gave tacit approval for Greek oversight of Smyrna and "the immediate area." The Greek advancements proved to be aggressive, as shown in Figure 2.5.[12]

FIGURE 2.5 Megali Idea, after the Treaty of Sèvres. Courtesy of the Creative Commons.

In spring of 1919, in consultation with Greek and British authorities, Toynbee helped to determine boundaries for what he understood would be a temporary Greek presence in Smyrna.[13] Butler, too, was consulted, having provided strategic advice to the British Office regarding the detailed maps he made while doing archaeological survey in Syria and his knowledge of Smyrna and Sardis.[14] It is also good to be reminded that Venizelos himself was at the conference, and his charismatic personality helped him spin things to his advantage. What is more, the Peace Commission did not send a group to Smyrna to assess the conditions on the ground, as it had done for Syria under the King-Crane Commission.[15]

For western Turkey, and specifically Sardis, the territory to be occupied was presented in Article 66 of Sèvres.[16] This included Smyrna, the Çeşme and Karaburun peninsulas, as well as the adjacent fertile river valleys. This zone encompassed classical archaeological sites and landscapes: Ephesus, Philadelphia, Sardis, Bin Tepe, the Gygaean Lake, Pergamon, and Assos. In fact, this desired territory corresponded with intentions to realize the *Megali Idea* decades after its proposal by former Greek Prime Minister Ioannis Kolettis and King Otto. As Liakos tells us,

> The celebrated text on the Great Idea of the Nation (Megali Idea) written in 1844 as an official parliamentary speech, referred to the psychic unity of

> the Greeks of the Greek Kingdom with the Greeks of the Ottoman Empire and to the contribution of Greek Civilisation to Western Civilisation, but not to territories to be claimed. Although all Greek parties and intellectuals supported the enlargement of the territory of the state, the image of Greece as a classical land prioritised Athens as the capital and the Greek peninsula as the national territory. On the other hand, the image of Greece as a continuation of the Byzantine Empire fuelled Greek ambitions to "restore" the Empire in Macedonia and Asia Minor and to acquire Constantinople as the future capital.[17]

This agenda was aided by the classical tradition in British, French, and American educational spheres. Many gentlemen of "good breeding" went on to be diplomats and businessmen in the Mediterranean, and patrons of archaeological expeditions.[18]

Among the most strident supporters of the Greek occupation was George Horton, then serving his second term as US Consul in Smyrna (1911–1917, 1919–1922). It was no secret that he was strongly in favor of Greek authority in western Anatolia.[19] His knowledge of the region and understanding of politics and diplomatic practice stemmed from his prior posts as US Consul in Athens (1893–1898, 1905–1906) and Salonika (1910–1911).[20] Horton's commitment to philhellenism included an aggressive vision to establish a strong US footprint in the region. He lobbied hard for US investors and companies, including the American Tobacco Company, McAndrews and Forbes, Socony (branch of the America Standard Oil Company of New York), Singer Sewing Machine Company, American Trade Development Corporation, and International Harvester.[21]

Among the key outcomes of the 1919 Paris Conference was the establishment of the League of Nations on January 10, 1920. The end goal, however, was never achieved. Even though the Treaty of Sèvres was signed on August 10, 1920, it was never ratified by Turkey. With the establishment of the Republic of Turkey, a new agreement was necessary: the Treaty of Lausanne, which was signed July 24, 1923. This entire process, from 1919 to 1923, reflects a period of time when unprecedented and drastic changes took place in the eastern Mediterranean and in the Near East. They had significant impact on negotiations of an international commission to oversee archaeology.

It is for this reason that I now turn to a detailed look at Sardis as a place that became entangled in webs of power, prestige, and philanthropy. The events that took place as a result of Sèvres put those committed to Sardis, especially those active in fieldwork and/or living in the area, in conflict with each other and presented great challenges for negotiations with authorities. This was made all the more challenging by the shifting of sovereignty in the region owing to the Greek occupation in May 1919 (i.e., territory claimed under Article 66 of the Treaty of Sèvres) and the Turkish War of Independence.

Networks of Sardis

Education merged with political power and economic standing in Princeton's American Society for the Excavation of Sardis. Members of the Sardis Society, listed in Table 2.1, represented a mosaic of shared educational experiences (i.e., Princeton, Yale, and Harvard), cultural connections, political aspirations, and business agendas. Displays of the World Fairs / Expos in London (1851, 1862), New York (1853), Philadelphia (1876), Chicago (1893), Paris (1889, 1900), Buffalo (1901), St. Louis (1904), and San Francisco (1915) between 1851 and 1904 broadened an interest in antiquity and the Orient in Europe and the United States.[22] The founding missions of the Metropolitan Museum of Art (MET) and the Boston Museum of Fine Arts (MFA) in 1870, followed by the Art Institute of Chicago in 1879, had exposed the US public to Mediterranean cultural history.[23] The connection between high culture and museology became intertwined with support for archaeology. In fact, among the ways to show continued commitment to the importance of the classical tradition and education was through financial support of expeditions.[24] In this way, archaeology benefited from late nineteenth- and early twentieth-century US economic expansion into Smyrna and philanthropic support of fieldwork, suggested by the leading title chosen for the October 2, 1910, *New York Times* article, "Home of Croesus, World's First Great Financier, Found."

To realize the Sardis expedition, the Society's executive committee and founding members included individuals with dual ties to business and the arts: MET director Edward Robinson; Lloyd Warren, founder of the Beaux-Arts Institute of Design in New York; Charles Munn, editor of *Scientific American*; Alison Armour, the wealthy Chicagoan with a passion for antiquity, national history, and exploration via his yacht, the *Utowana*; and the oil-fortune heir V. Everit Macy. It also included the young and wealthy Robert Garrett, who while at Princeton competed in the first "modern" Olympics of 1896 in Greece, taking home the gold medal for discus throwing. What is more, the banking tycoon and avid collector John Pierpont Morgan, too, supported the expedition. Well-known architect Thomas Hastings[25] served as the Executive Committee Chairman.

TABLE 2.1 Members of the American Society for the Excavation of Sardis

AMERICAN SOCIETY FOR THE EXCAVATION OF SARDIS

MEMBERSHIP FOR THE YEAR 1921

Executive Committee

Mr. Thomas Hastings, *Chairman*
Mrs. James A. Burden Jr.
Miss Josephine Osborn
Mr. Edward Robinson
Mr. Lloyd Warren, *Secretary*
Mr. Allison V. Armour
Mr. Charles A. Munn
Mr. V. Everit Macy, *Treasurer*
Professor Howard Crosby Butler, *Director of Excavations*

Founders

Mr. Allison V. Armour
Mr. William H. Buckler
Mr. Robert Garrett
Mr. Edward S. Harkness
Mrs. Edward S. Harkness
Mr. Archer M. Huntington
Mr. V. Everit Macy

Fellows

Mrs. Charles B. Alexander
Mrs. James A. Burden Jr.
Miss Agnes Miles Carpenter
Mr. Cleveland H. Dodge
Mr. Alessandro Fabbri
Mr. Ernesto G. Fabbri
Mrs. William P. Hamilton
Mr. Cyrus H. McCormick
Mr. Charles A. Munn
Mr. Louis C. Tiffany
Col. Robert M. Thompson
Mr. Lloyd Warren
Mr. Francis M. Whitehouse
Mrs. Francis M. Whitehouse

Sustaining Members

Miss Eleanor de Graff Cuyler
Mr. Thomas Hastings
Mrs. Walter B. James
Mr. Gerard B. Lambert
Mr. Walter E. Maynard
Prof. Henry Fairfield Osborn
Mrs. Henry Fairfield Osborn
Mr. William Church Osborn
Mrs. William Church Osborn
Mrs. Herbert L. Satterlee

Annual Members

Mrs. Gerard B. Lambert
Prof. A. Kingsley Porter
Mr. Edward Robinson
Mrs. Edward Robinson
Dr. T. Leslie Shear
Mr. Roderick Terry Jr.

DECEASED MEMBERS

Founder

Mr. J. Pierpont Morgan

Fellows

Mr. John L. Cadwalader
Col. Henry L. Higginson
Mrs. Mary J. Kingsland
Mr. M. Taylor Pyne
Mrs. Archibald D. Russell

Sustaining Members

Mrs. Lloyd Brice
Mr. Frederick W. Whitridge
Mr. Arthur P. Sturges

The cultural dimension of the full board complemented the executive committee: the Osborn and Sturges families, with their unfailing loyalty to the MET; the famous design director Louis Tiffany of Tiffany & Company; and Mrs. William P. Hamilton, who had become accustomed to the arts and collecting through her father, J. P. Morgan. Expertise in architectural studies, too, pervaded the board. In addition to Hastings and Warren, architect Francis M. Whitehouse bolstered Sardis' reputation in this sphere.

Others came with experience in politics, such as John Lambert Cadwalader, whose broad interests were shown in his appointments as trustee for both the Carnegie Endowment for International Peace and the MET. He was close with the spirited Col. Henry L. Higginson, founder of the Boston Symphony Orchestra. What is more, both Cadwalader and Higginson knew foreign policy expert Cleveland H. Dodge. These men, and especially Dodge, were trusted members of US President Wilson's inner circle. In fact, scholars describe Dodge as the most important diplomatic operative for US foreign policy in the Middle East.[26]

Business interests, too, brought people together. The trade in opium had drawn Russell and Company, headquartered in Connecticut, to Smyrna.[27] Eventually, Mrs. Archibald D. Russell patronized Sardis, as did the Sturgis and Huntington families, who had also invested in the opium business.[28] The expanding production of opium poppies led to Gerard B. Lambert's interests in the region and, thus, his patronage of Sardis. His St. Louis company, Lambert Pharmacal, eventually became very successful under the Listerine® brand, only recently bought by Pfizer.

Whether for the cultivation of opium, tobacco, or sultanas, mechanical farming was another expanding opportunity. Sardis fellow Cyrus Hall McCormick, Jr., had personal interest in Sardis; his son, Gordon, trained as an architect during the 1922 Sardis season (see later in this chapter). Yet the family also had business aspirations for International Harvester, which in 1902 replaced McCormick Harvesting Machine. Working with Horton in Smyrna, McCormick hoped to convince farmers in the region to try new approaches.[29] On September 25, 1919, the *American Machinist* published this statement by Horton, vol. 51, No. 13, 640.

> A farm of 20,000 acres, on a railroad 17 miles from Smyrna, is owned by an American citizen. Part of this farm could be utilized as an experiment station for tractor and other machinery. . . . The Greek government informs me that it is willing to pay for tractor machines sent over for introductory purposes; and also that it will place an order for farm machinery to be used

> by Turkish and other refugees when they return to their farms. Samples suitable for this soil are desired by the government.

US companies worked hard to cultivate an image of adaptability in order to secure entry into Smyrna's markets. Calculated, shrewd business tactics are revealed in an essay that appeared on April 6, 1914, in the *Chicago Daily Tribune*:

> Such work takes time. American success in the Levantine market to date has been gained by studying oriental peculiarities, by patience, by adaptation, and by employing men and methods demanded by Turkish conditions. Competition is stiff . . . Once introduced there is the same devotion to a favorite line, the same unwillingness to change, as shown by the Arabs in Egypt. (Frothingham, 1914, p. 13)

In 1910 the American Standard Oil Company of New York had a limited footprint in Smyrna. By 1919 the company had expanded operations considerably, including storage docks and a factory for the production of tins and cases for oil.[30] Sardis Society members whose family wealth derived from Standard Oil included Mr. and Mrs. Harkness, also patrons of the MET. Olympian Robert Garrett, too, owed his fortune to black gold, railroads, and finance.[31]

This network of prominent people who supported Sardis were not alone in their interest of archaeology and objects from the region. The collections of Edward Perry Warren (Warren Bros. & Company of Boston, Electric Power Engineers) and Alfred O. van Lennep (Dutch Vice Consul in Smyrna) had raised eyebrows. Established institutions (e.g., the British Museum and Louvre) purchased items from Smyrna, as did US institutions: the MFA, the MET, the Oriental Institute, Harvard University, and the University of Pennsylvania. Archaeologists with the AIA functioned as an informal consulting group for museums and private collectors. Institutions were savvy in their approach with collectors as well as with US customs and the US tax code.[32] Coins, jewelry (e.g., necklaces, rings, and gems), and terracotta and stone sculptures were among Warren's favorites, which he donated to the MFA Boston. Several items (e.g., gold staters) are attributed to Sardis specifically. Items donated to the Louvre, too, are attributed to the region, including lions (i.e., Lydian Lion). What is more, objects given to the British Museum by George Dennis include Roman lamps from Bin Tepe (likely indications of Roman-era plunder), fragments of furniture, perfume jars, and ceramic vessels. Records show also that the British Museum purchased (from Dennis) two sculpted reliefs from Bin Tepe (18891021.1 and 1889 1021.2).

For Sardis Society members, the Sardis Expedition was a logical "next step" to expand the philhellenic footprint into western Anatolia and to secure more objects for US museums. The relationship with Hamdi Bey, Director of the Imperial Museum in Constantinople, and the diplomatic savvy of Armour, Buckler, and Garrett secured the Ottoman permit (*firman*) for the expedition, which was ultimately, according to Butler, vested with the American Embassy.[33] To be sure, however, it was the work of Butler in Syria that had resulted in the Ottoman authorities inviting him to excavate at Sardis.[34] In preparing the application, they had to edge out the interests of Professor Kelsey at the University of Michigan.[35] Among the reasons that Michigan could not be a serious contender was that those supporting the expedition wanted the material on the east coast, preferably in Baltimore or at the MET.

The MET and the cultural elite of New York City anticipated that they would receive objects as part of the gentleman's agreement of *partage* between Buckler, Garrett, Armour, and Hamdi Bey. Robinson, Morgan, and Macy shared this position. Macy wrote to Butler,

> Enclosed is a copy of a letter from Mr. Robinson in regard to Mr. Morgan's subscription for the coming year. From the tone of the letter you will see that it will be most important to get out as many of the finds to America during the coming year as possible. Can you not follow the policy of having everyone who leaves Sardis, whether they are visitors or not, take some "fragments" with them? I am afraid that if you wait until you leave, and should then attempt to take out several boxes of stuff, you might be held up. As I have told you from the start, the donors are more interested in what can be obtained for the Museum than what is found. Of course, if objects of great value are discovered, the donors, I do not suppose, will expect to receive all, but if the Turkish Government, on the other hand, insists on keeping all, I do not believe the subscriptions would be continued."[36]

In essence, Sardis represented an untapped mine, a site of extraction. Macy referred to finds of nice objects as "strikes" and "spoils."[37] And his support depended on sustained returns on investment. He wrote,

> I am free to confess that unless the Metropolitan Museum is going to receive something in return for the money that has been subscribed, I would not

care to renew my subscription. I am not sufficiently interested in the scientific value of excavation to continue my contribution. You see, at heart I am really, very much of a Philhellene.[38]

This put pressure on Butler to excavate in areas likely to produce the desired objects and to negotiate with the Ottoman authorities for export permission. The alternative was willful blindness to informal smuggling operations. This disjuncture between the collecting elite of New York, Boston, and Smyrna, and scholars in the field, such as Butler, led to tensions in balancing Sardis' allegiance and dependence on the weakening Ottoman regime.

The Sardis field team included a few board members, such as Buckler (and others, such as Robinson, visited briefly) as well as students of archaeology from Princeton. Butler imported not only his field tents and equipment from the Levant, but also three Christian Arab house staff, who had worked with him in Syria. The Turkish commissioner (government representative) was a Sephardic Jew from Smyrna. The artisans who built the dig house Villa Omphale were "Cretans."[39] Local Muslim men worked as day labors in the excavation.

In less than a few months, the excavation team transformed the Temple of Artemis into a full-scale site of extraction. As depicted in Figure 2.6 the Temple foundations were buried at the start of the Butler expedition; only two columns indicated the former glory. As revealed in Figure 2.7, the Temple emerged from the Anatolian soil. From the top of the Necropolis, from where the photos were taken, the Temple stood naked. She was flanked by the spoils of the industrial operation: back-dirt, a crane, trail track, and rail cars, as shown in Figures 2.8 and 2.9. The excessive quantities of dirt created a terrace between the Temple and the Pactolus, and thus alleviated fears of erosion from the Pactolus encroaching on the Temple. Despite the rawness of the mechanized excavation process, subsequent illustrations of the monuments were prepared in exquisite detail following the *Beaux Arts* style, as shown in Figure 2.10 of a column capital.[40]

In 1914 the expedition was put on hold. It was both the threat of war and later the reality of it that pushed Butler and Buckler to consider management of antiquities. Between 1917 and 1919, they tightened their convictions to establish an international commission. Yet, as tensions between Greece and Turkey increased, and the Turkish War of Independence erupted, those involved with Sardis blurred the boundaries of policy, practice, and law. Correspondences between Buckler and Butler and their colleagues reveal calculated choices to defy regulations and Ottoman authorities.

FIGURE 2.6 Snow on the ruins of the Temple of Artemis at Sardis (prior to excavation). No. A.245. American Society for the Excavation of Sardis 1910–1914. Courtesy of the Department of Art and Archaeology, Princeton University.

The Treaty of Sèvres

In December 1918, Butler proposed that the AIA appoint Buckler "as a sort of special agent to reach the Peace Commission."[41] This plan worked. In 1919 Buckler pushed forward in Paris to implement Article 421 of Sèvres:

FIGURE 2.7 The extraction of the Temple of Artemis at Sardis, side view. No. A.256. American Society for the Excavation of Sardis 1910–1914. Courtesy of Department of Art and Archaeology, Princeton University.

The Turkish Government will, within twelve months from the coming into force of the present Treaty, abrogate the existing law of antiquities and take the necessary steps to enact a new law of antiquities which will be based on the rules contained in the Annex hereto, and must be submitted to the Financial Commission for approval before being submitted to the Turkish

FIGURE 2.8 Crane moving capital at Sardis. No. A.118.b. American Society for the Excavation of Sardis 1910–1914. Courtesy of the Department of Art and Archaeology, Princeton University.

FIGURE 2.9 Rail system for backdirt at Sardis. No. B. 312. American Society for the Excavation of Sardis 1910–1914. Courtesy of the Department of Art and Archaeology, Princeton University.

FIGURE 2.10 Detail of the column capital from Sardis. No. A262-1.American Society for the Excavation of Sardis 1910–1914. Courtesy of the Department of Art and Archaeology, Princeton University.

> Parliament. The Turkish Government undertakes to ensure the execution of this law on a basis of perfect equality between all nations.

The devil was in the details of eight points outlined in the Annex, with the last two speaking directly to the Allied agenda. Point seven addressed the rights of foreign excavations in Turkey; point eight outlined terms for *partage*. The final line for point eight is worthy of careful consideration: "If division seems impossible for scientific reasons, the excavator shall receive a *fair indemnity in lieu of a part of the find*" (emphasis added).[42] Archaeology, apparently, was befitting of a business deal, akin to concessions for railways, ports, or mines. In this way, the monetary arrangements of *partage*, or equivalent payment, were woven into the rhetoric of the global good and the practice of classical archaeology.

Butler himself promised Breasted and Robinson that among the reasons to pursue the new legislation at an international level was "to put an end to the systematic neglect and destruction of historic monuments, to the commercialization of antiquities, and to the obstruction of legitimate scientific exploration and excavation, which have hitherto prevailed in countries under Turkish rule."[43] As discussed earlier in this chapter, the Sardis Society funders professed their disinterest in science, their love of objects, and their desire for them to be owned by US institutions, whether through legal means or unofficial policies to secure items through what amounted to smuggling operations. To his credit, Butler attempted to pursue a policy suggestive of greater ethical intent, even though he found himself in an impossible situation with conflicting interests.

This narrative built on the revolving legacy of US archaeological practice and infrastructure. The AIA, first established in 1879, assisted with the founding of the ASCSA in 1882, conducted its first major international restoration project at Assos (Turkey) between 1881 and 1883, and assisted in the establishment of the American School of Classical Studies in Rome in 1895 and the opening of the American School for Oriental Study and Research in Jerusalem in 1899 (the precursor to ASOR). These actions provided evidence for the AIA's relevance to US foreign policy, as presented in the 1906 AIA's US Congressional Charter:

> [P]romoting archaeological studies by investigation and research in the United States and foreign countries by sending out expeditions for special investigation, by aiding the efforts of the independent explorers, by publication of archaeological papers, and reports of the results of the expeditions which the Institute may undertake or promote, and by any other means which may from time to time be desirable.

Section 3 of the charter made financial provisions such that the AIA could "hold real estate and personal property in the United States and any foreign country for the necessary use and purposes of said organization."[44] No specific provisions were given for the monetary value of objects, however, nor did AIA President Charles Eliot Norton connect archaeology to the art market in his 1900 publication for the *American Journal of Archaeology*, "The Work of the Archaeological Institute of America: An Address."

To the contrary, Norton waxed eloquently of scientific endeavors, the value of the humanities to the human spirit, and the importance of both in the fight against the lure of materialism.[45] He argued that the great wealth of the late nineteenth century clouded the focus of scholarship and belittled the importance of education, explicitly the humanities:

> At such a period as this, the need is great that those who prize the humanities as the strongest forces in the never-ending contest against the degrading influences of the spirit of materialism, as the best means of development and discipline of the intelligence, as the source of the knowledge most useful for the invigoration and elevation of character, and most abundant in nutriment for the noblest intellectual qualities, — the need is great, I say, for

those who hold the humanities in this esteem, and above all for those who recognize in classical studies, largely interpreted and rightly understood, the quintessence of the humanities, to unite in the assertion and maintenance of the supremacy of these studies among the general elements of the higher education. To this end the Institute and its Schools have contributed."[46]

Norton's assessment that the AIA and her partner foreign schools in Athens, Rome, and Jerusalem had influenced classical education was certainly true, but so, too, was it true that their approach had significant impact on US foreign policy and international diplomacy.

In 1911 an AIA office had been established in Washington, D.C. In the eyes of the AIA and her members, the Paris Peace Conference offered an opportunity to make a difference and benefit its membership. Between 1917 and 1919 formal reports, newspaper articles, and personal correspondences focused on questions regarding the sovereignty of archaeological sites and associated monuments and objects. Butler and Buckler unveiled their strategy for the AIA in 1917. Butler put forth a resolution, which was adopted, that the AIA "appoint a committee to communicate with the government in Washington on the subject of the protection of the historic monuments in Turkey as soon as peace negotiations should be begun." On January 2, 1918, General Secretary, G. M. Whicher of the AIA wrote to Butler that the AIA would move forward with deliberations. A year later, in December 1918, the AIA voted favorably.[47]

Butler received word from Buckler (then at the US Embassy in London) of a newly established group of stakeholder institutions under the umbrella of the Hellenic Council, as listed in Table 2.2. Two letters (dated February 7 and March 11, 1919) gave Butler and Buckler the support they needed from fourteen US organizations, shown in Table 2.3.[48] The momentum also assisted with the establishment of the US Institute of International Education and the American Council of Learned Societies (ACLS), both founded in 1919.[49] Funding for the ACLS came from the Carnegie Endowment, and it is worthy of note that John Lambert Cadwalader was a trustee for this organization as well as a member of the Sardis Society. He and others from the "Eastern Establishment" propelled significant support for archaeology and its integration with academic institutions, including those in Britain and France.[50]

Buckler leveraged the British Hellenic Council, the French Union Académique Internationale (International Union of Academies), and British and French diplomats to convince further the American delegation of the AIA's initiative at the Conference. This gave Buckler "proof" of

TABLE 2.2 Members of the Hellenic Council

The British Museum
The Victoria & Albert Museum
The Societies of Antiquities, of Hellenic Studies, and of Roman Studies
The Royal Institute of British Architects
The Royal Anthropological Institute
The Royal Asiatic and Numismatic Societies
The British Schools at Athens and Rome [and the future in Jerusalem]
The Egypt Exploration Fund
The Palestine Exploration Fund
The Byzantine Fund
The Institute of Archaeology in the University of Liverpool.[101]

TABLE 2.3 U.S. signees of the AIA initiative at the 1919 Paris Peace Conference

The Archaeological Institute of America
The American Academy in Rome
The American Geographical Society
The American Historical Association
The American Numismatic Society
The American Oriental Society
The American Philological Association
The Art Institute of Chicago
The Boston Museum of Fine Arts
The Field Museum of Natural History, Chicago
The Haskell Oriental Museum and Department of Oriental Languages of University of Chicago
The Metropolitan Museum, New York
The National Institute of Arts and Letters
The Toledo Museum of Art

interest, of importance. In fact, these actions represented an important moment not only in the history of archaeology and international legislation, but also in the rallying call for learned societies, including the Institute of International Affairs, initiated by the British and American delegations at the Conference. The result was a common set of cultural codes and growing power of corporate philanthropy.[51]

With the backing of key groups, in March 1919 Buckler (in Paris) pursued an "Internationalization of the Historic Monuments of Nearer Asia."[52] He was joined by Salomon Reinach and Rene Cagnat of the Institute of France, Lieutenant Commander Hogarth of the British Academy, and Roberto Paribeni, Director of the National Museum in Rome. This group of five acted quickly:

> [As the Covenant of the League of Nations] would shortly be adopted by the Peace Conference and negotiations for the settlement of the Ottoman Empire might soon afterwards begin, it was desirable immediately to draft proposals for submission to the Conference. To have waited for full discussion by the academies of the Allied and Associated Powers, and for the devising by those bodies of measures to attain the object in view, would have involved so much delay as perhaps to defeat that objective.[53]

They outlined the organization of a future "Commission to Regulate Mandates under the League of Nations." They envisioned an international observatory that would establish a Sub-Commission on History and Archaeology.[54] This Sub-Commission would include nine people, of whom eight would be archaeologists. In collaboration with the Council of the League, the Commission on Mandates would make appointments of five-year terms. The Commission would receive reports from the Mandatory Powers concerning the condition and stewardship of archaeology within their respective boundaries.

On June 3, 1919, the Sardis Society Executive Committee met to review a proposal to be presented at the Peace Conference. The AIA membership learned of the initiative in the September 1919 issue of the *Art Bulletin*:

> The collapse of the Turkish Empire has called the attention of the civilized world to the importance of protecting the ancient historic monuments and objects of art which for centuries have been under the careless rule of a government that has had little or no interest in them. No lands on the globe contain such rich treasures of antiquity, occupying so vast an area, representing so many civilizations, and covering so long a period of the world's history. Most of the early history of our own civilization and art lies buried in these lands which are now to be placed under some form of control by the leading powers of the western world. It is manifestly the duty of these powers to take immediate steps to protect this ancient heritage—of which, after all, the western world is the true heir—and to formulate laws, and make common agreements, according to which the historic monuments and the works of ancient art now buried may be brought to light and made most efficiently to serve the demands of civilization.[55]

Butler conveyed also the desire for the sale of antiquities by dealers and private individuals in order to "guarantee protection of the antiquities from loss or damage."[56]

FIGURE 2.11 Council of Four at the WWI Paris Peace Conference, May 27, 1919 (candid photo) (L–R) Prime Minister David Lloyd George (Great Britain); Premier Vittorio Orlando, Italy; French Premier Georges Clemenceau; US President Woodrow Wilson. Photographer, Edward N. Jackson (US Army Signal Corps). Courtesy of Creative Commons.

A month later, on October 18, 1919, proposals to control heritage in Turkey as well as the sale of it were submitted for review to a meeting of the Union Académique Internationale. Delegates from the Peace conference attended and gave their unanimous endorsement, despite that "no delegate had been authorized to deal with this subject."[57] With this decision, Buckler had secured the blessing of Sir Eric Drummond, Secretary-General of the League of Nations, and thus the ability to move forward with integrating specific language for access to archaeological sites and antiquities in Turkey into Sèvres. The big four—British, Italians, French, and the Americans—endorsed the proposal (Figure 2.11).

Smyrna's Countryside and the League of Nations

During the exact period—spring 1919—when delegates in Paris focused on the provisions of an international commission for antiquities, Smyrna and her countryside were occupied by Greek military forces—with the blessing of the British, and the tacit approval of the other Allied members. Following the Greek occupation of Smyrna in May 1919, the Greek Archaeological Service in consultation with ASCSA was given authority over archaeological matters. Questions arose about Sardis, ca. 85 kilometers east of Smyrna.[58] Recall that the boundary between Turkish and Greek territory was a wire fence that ran up the Acropolis. Many argued that Ottoman officials were neglectful and that Greece and the ASCSA would offer greater security for antiquities. The reality of Ottoman neglect, however, is debated.[59]

By the mid-nineteenth century, Ottoman authorities were sufficiently frustrated and burdened by ambitious foreign expeditions meddling in legal parameters of antiquities and monuments in their jurisdiction, especially in the region of Ephesus, 80 kilometers south of Smyrna. Edhem Eldem gives a rich picture of the exasperation of the Public Works Commission of the Council of State with the European pursuit of antiquities. A March 1869 law allowed for export permission pending diplomatic correspondence between the Sultan and the respective state. Updates were made in 1874, followed by formal regulations regarding ownership in the 1884 Antiquities Law.[60] Despite this legal structure, foreign archaeologists and their expeditions continued to push forward with desires for gentlemen *partage* arrangements. Horton, too, vehemently supported these measures, and helped to usher in the ASCSA as the overseer of work at the archaeological site of Kolophon (40 kilometers south of Smyrna) during the Greek occupation of Smyrna.[61]

This group presumed that the conditions under the Treaty of Sèvres (i.e., Greek control of Smyrna's hinterland) were faits accomplis. The minutes of the ASCSA meeting of May 10, 1919—a mere five days prior to the Greek landing in Smyrna—made no mention of this very significant and impending event. Even when the secret agreements about Smyrna between Venizelos and Lloyd George were revealed (i.e., the British promise to the Greeks to edge out the Italians), Buckler pursued a mechanism that would align archaeology and business assets, and Butler explored how to evade Turkish law in order to export material from Sardis.[62] Buckler attempted to woo support from US Secretary of State Robert Lansing. To be sure, Lansing's attention extended well beyond archaeology. He wanted to know the potential impact on existing US arrangements (i.e., trade agreements) with Ottoman authorities, should political authority significantly shift.[63] In addition to future presumptions about Sèvres and Smyrna, many in Europe operated under the assumption that the United States would be a member of the League of Nations and take a mandate for Turkey. Buckler favored this position.

In late November 1919 Buckler, on behalf of the US delegation, drafted and sent proposals to Washington.[64] He outlined the primary positions of the British, French, and Italian delegations, and ongoing Greek aggressions in western Anatolia. Buckler advocated for US involvement on grounds of "National Safety and Protection to National Interests." US engagement was critical for the security of Turkey and without it, Buckler stressed, the United States would "gravely jeopardize American moral prestige, American religious and educational interests, and the large and varied American business interests in the Near East." The signees included prominent groups in Turkey, as listed in Table 2.4. All feared the consequences to their respective interests should multinational diplomacy collapse.

The League of Nations was promulgated on January 10, 1920, yet without US endorsement. On August 10, 1920, the British, French, Italian, Japanese, and the Ottoman Empire signed the Treaty of Sèvres. During the 1921 winter conference in Paris (February–March), the Greek and Turkish delegations were unable to agree on the future sovereignty of Smyrna, Constantinople, and eastern Thrace.[65] Growing at the same time were Greek ambitions (with British approval) to move well beyond the boundaries of Sèvres. In fact, the line at Sardis had been crossed in June 1920. Despite being a member of the League of Nations, and thus subject to Article 10 regarding human integrity, Greek military forces pursued a brutal military campaign in western Anatolia.

In his letter of November 4, 1919, Venizelos had stood his ground. He argued that Greece was only doing what her allies wanted, yet didn't have the spine to carry out:

TABLE 2.4 Signees of Buckler's Memorandum on the policy of the United States Relative to the Treaty with Turkey, 26 November 1919.

Board of Foreign Missions
American Committee for Relief in the Near East
President of Roberts College
Director of the American College for Girls at Constantinople
The Standard Oil Co. of New York
McAndrews and Forbes
The American Tobacco Company
The American Foreign Trading Co.
The Singer Sewing Machine Co.

> [T]he Greek Government complies with the real intentions of the Allied Powers, which for these reasons, should instruct their agents at Smyrna to consider the situation with the objectivity necessary in order to avoid the regrettable complications which the enemies of the Entente alone are able to take advantage of.[66]

According to Buckler's assessment, Venizelos gauged the fear of the US, British, French, and Italian delegations should Smyrna be subsumed within Turkish territory. Venizelos gambled that the Great Powers would turn a blind eye to Greek military advancements inland to preserve their own economic arrangements. Access to antiquity became part of the deal. Ultimately, to Buckler's disappointment, the United States didn't join the League (in order to preserve US trade interests and US isolationism). The Treaty of Sèvres crumbled, and with it the specifics on antiquities.

After the US withdrawal from the League of Nations in early 1920, the ACLS met at the Institute of International Education on February 14, 1920. W. G. Leland proposed "protection of historical and archaeological interests in Western Asia."[67] Between 1920 and 1922, authorities in Smyrna and Istanbul, as well as those in nearby Salihli, focused on security at Sardis. In addition, Butler and Buckler worked closely with members of the Sardis Society to raise funds and to arrange for those in Smyrna to make periodic trips to assess the conditions at the site. They also made arrangements to have field equipment, archives of the excavations, and antiquities taken from the site and shipped to New York, contrary to what they told Ottoman authorities.

Post-Paris: Crisis at Sardis

A contrived ignorance justified inaction to combat the indignities to local communities enacted by Greek military forces in the hinterland of Smyrna. It was not only the reluctance to acknowledge, but also the refusal to act on Turkish reports of the "horrors and crimes" that had begun with the landing of Greek forces and their steady advancement inward in May 1919. The blatant disregard of the right to human dignity in Article 10 of the League of Nations put in motion a culture of avoidance whereby those in Paris ostensibly found the capacity to ignore the reports of the atrocities committed by military forces in Smyrna's hub and countryside. Pursuit of territory in the name of protecting heritage, regardless of the impact on the immediate populations, took precedent.[68]

Buckler's correspondences reveal his personal struggle. On May 23, 1919, Buckler wrote to his wife Georgina that it was "appalling to think how many far-reaching decisions are being taken by a few men who know but little about their subject." Even so, he rejoiced in having been present "when the Turkish die was being caste."[69] On June 5, 1919, he reported that he and others were "trying to devise . . . some way of stopping the massacres produced in Asia Minor by the Greek invasion."[70] Recall that several months earlier, on March 31, 1919, Buckler deliberated, "To my mind, the reconstruction of Turkey calls for a fine bit of real statesmanship. . . . Every citizen of that country should be guaranteed under the authority of the Powers, absolute security of property, life and religion." [71] Despite these considerations, there is no evidence (as of yet) that he acted with explicit purpose to garner support to quash Venizelos or his forces.

Reports arrived from Smyrna with thick descriptions of "towns, villages and hamlets [turned into] heaps of ruins and ashes" including the "smoking wreckage, the carbonized corpses, the bleeding remains of thousands of poor innocent Turks, of women, of children, of old men sacrificed to the ferocity of the Hellenic hordes."[72] Still, Venizelos was not expelled from Paris. This despite more evidence from Ottoman officials of the blatant and horrific attacks on human rights. The "systematic destruction" resulted in such bloodshed that communities in Smyrna claimed "under whatever regime we may live our dead will always have a hold over Hellenism."[73]

Smyrna's wealthy, powerful, and influential Levantine communities further substantiated these claims.[74] In their October 20, 1919, document, "Appeal to the Great Powers and to the World's Opinion in Favour of the Refugees of Smyrna," they called the Allied Powers to task:

> Are the foreigners more inclined to pass from Ottoman "barbarism" to Greek "civilisation"? Not at all. . . . The English Chamber of Commerce at Smyrna has taken the decision to do its utmost to prevent Smyrna's being separated from Asia Minor. Thus the families, heads of their respective colonies, the Guiffret and Girods of France as well as the Whitals of England and the Grynns of America. And passing from wishes to actions, and from words to deeds, they have sent delegates to Paris and to London, to demand a thing so simple and yet it seems, so complicated in these days of the peoples' right to self-determination that Turkey should remain in the possession of her children, and to obtain the solution which alone seems to them just and profitable, the maintenance of Turkey. . . . Thus on the rights and the deeds in the question of Anatolia we demand the complete publication of the reports of the Allied Inquiry Commission at Smyrna.[75]

Members in Paris finally reprimanded Venizelos. He was told to get ahold of his men in the field. Yet, his military wasn't stopped. The carnage continued.

The Sardis Society reasoned that the Allies would eventually hold power in not only Istanbul but also Smyrna. In fact, warnings had arrived a year earlier. These came from William Ramsay Berry, a colleague living Smyrna, who had written the following to Butler on April 27, 1918:

> The forces which are at work sweep along with them in their gigantic motion all human beings and no person exercises any real influence over the conduct of affairs, with the possible exception of President Wilson, and it remains to be seen whether he makes a success of it. Never has such enormous power been concentrated in the hands of any living being.[76]

The United States had taken on more power, more prominence, and yet many questioned American fortitude and understanding of the situation.

With this in mind, planning at Sardis had begun in summer 1918, yet not only by the Americans, as shown in Table 2.5. Ottoman officials, too, had anticipated future tensions. Major Ernst Rosenwaldt visited the area in April and May of 1918; his report was sent to Halil Edhem, Director General of the Ottoman Imperial Museum in Constantinople, who had taken over the post following the death of his brother, Osman Hamdi Bey. Edhem also received reports from Salih Zeki Bey, Director of Public Instruction of the Smyrna Vilayet, who visited Sardis in May 1918 upon orders from Izzet Bey, Minister of Public Instruction. Given these two reports, Halil requested that Georg Karo, director of the German Archaeological Institute of Athens, and fellow archaeologist H. Knackfuss visit Sardis to make a full assessment. On June 21, 1918, they did so. A copy of their report was given to the Governor General of Smyrna.[77]

At Sardis, they found that while Villa Omphale had been entered, the damage was minimal. Precautions were taken to secure valuable objects and to pack (in three trunks) excavation notes, photographs, instruments, and rare books. According to Karo, only the trunk with the "private effects of the learned Americans, and of which Sahil Zeki Bey ha[d] the key" was to be transported to the Swedish Consulate in Smyrna; the other two trunks with artifacts were to be transported to the Mekteb-i Sultani in Konak for safekeeping. Karo and Knackfuss credit Sahil Bey and Izzet Bey with facilitating the operation. On July 24, 1918, Swedish Consul Van der Zee wrote to the US Charge d'Affairs regarding the export of "the American"

TABLE 2.5 Key personnel involved in the in-country diplomacy at Sardis

William R. Berry	Captain with the US War Department and engineer in Smyrna
Halil Edhem	Director General of the Ottoman Imperial Museum in Istanbul
Lewis Heck	US High Commissioner in Istanbul 1915–1917
George Horton	US Consul in Athens 1893–1898 and 1905–1906; Consul General in Salonika 1910–1911; and US Consul in Smyrna 1911–1917; author of *The Blight of Asia*
Georg Karo	Director of the German Archaeological Institute of Athens
H. Knackfuss	German Archaeological Institute of Athens
S.A. Papafrango	Envoy Extraordinary of the Royal Government of Greece
Mr. Aristidis Sterghiades	Greek High Commissioner, Smyrna
Major Ernst Rosenwaldt	Medical doctor posted to Smyrna (focus on Malaria)
Alfred Van der Zee	Dutch-Levantine Consul for Denmark and Sweden in Smyrna
Salih Zeki Bey	Director of Public Instruction of the Smyrna Vilayet
Izzet Bey	Minister of Public Instruction

trunk. In addition, he clarified that Sahil Bey, in consultation with the Ministry of Public Instruction and the Ottoman Imperial Museums, had given public funds to pay for a guard at Sardis.[78]

Secretary of State Robert Lansing weighed in on security at Sardis on January 20, 1919, two days after the Paris Peace Conference had begun.[79] In his report of July 22, 1919, first Lieutenant Tefik, Commander of the Jandarma in Salihli, reported that Greek forces had plundered Sardis and kidnapped members of the Jandarma, forest keepers, the Sardis guard, and seven locals. Villa Omphale had been further pillaged, yet not the Temple of Artemis. The Jandarma's report concluded, "As Sardis is now a battle front it is not safe to leave there a gendarme or guard."[80]

On September 6, 1919, Lansing's office informed Butler that they had detailed Captain William Berry with the US War Department in Smyrna to visit the Sardis Expedition, only to rescind the offer on October 9, 1919.[81] The correspondences between Horton, Berry, Toynbee, and others point fingers at Turkish forces, claiming they looted Sardis between fall 1919 and winter 1921. Toynbee wrote on February 10, 1921, not only that the Turkish forces had destroyed the excavation house, but also that the "statues in the central court have been badly defaced; arms faces, etc., mutilated; the pottery in the big shed at back smashed to smithereens, but luckily the Lydian inscriptions are intact."[82]

Eventually, the Sardis Society asked (and funded) Berry to visit Sardis in March of 1921. Berry's detailed report addressed the conditions at Sardis and the sentiments of the vested parties. According to him, in spring of 1920 under command of Kemalist forces, "lawless Circassians"[83] had burned villages to prepare for the Greek military advancement across the Sèvres boundary (the Milne Line, named after the British general who was to enforce the boundary), which came in late June 1920. Berry reported that the facilities at Sardis, including the main house, museum/depot, and tennis court had been extensively plundered as of spring 1921, and even the expedition's horse (Baby) had suffered. Berry also included population data before and after the Greek military advancement, as shown in Table 2.6.

Yet, despite the traumascape that the numbers portrayed, Berry suggested the possibility of a field season for summer 1921. He claimed that the headquarters of the Fifth Greek Division in Salihli provided sufficient security. Archaeology would be good for the cause, he argued:

> [I]t would be to the advantage of the Greek cause in Asia Minor to hasten as much as possible the resumption of American excavation at Sardis, not only because thereby Greece could demonstrate her hospitality under trying circumstances towards scientific research, but also because the mere presence, secure and undisturbed, of a foreign archaeological expedition in the heart of the Greek occupied area would serve as a particularly convincing illustration to western eyes (at present so ill-informed) of the tranquil conditions established and maintained by a conquering and locally unpopular army.

According to Berry, Greek forces considered their jurisdiction permanent: "In the general Greek opinion, official and unofficial, as well as in my own, the Greek Government will not relax its hold on that part

TABLE 2.6 Population data before and after Greek occupation near Sart. Provided by W. R. Berry in his 1921 Report to the Sardis Society

VILLAGE	HOUSES BEFORE	HOUSES AFTER	POPULATION BEFORE	POPULATION AFTER
Sardis	14	6	60	12
Second village	17	4	80	19
Sardis Station	All houses, except station, destroyed and deserted.			
Uruk [Yörük]	Destroyed and deserted, former inhabitants in hiding in the hills.			

of Anatolia which it now occupies under any circumstances—short of pressure from popular opinion at home, or being driven out at the point of the sword."

There was concern of the symbolism embedded in the occupation of Smyrna. Berry described threats to regional security, as quoted in the outset of this chapter:

> There is a marked spirit of depression; last hopes are centered on the London Conference. Halil Bey asked if the Allies are bent upon the utter destruction of Turkey, for only thus can he interpret their attitude. It appears to be the common opinion that Turkey's chances for the future rest upon the ultimate disposition of Smyrna. With Smyrna a peaceful and national development of Turkey-in-Asia by the Turks and for the Turks is possible: without Smyrna, fruitless dissention and eventual disintegration are thought inevitable.

The terms of Sèvres may have died at the winter 1921 conference, but Greek military advancement continued, accelerating the conflict.[84] It was a house of cards. As much as the Allies wished to pivot away from "Greek influence and to withdraw foreign troops from Anatolia to appease the Turks, both failed to appreciate the extent to which their own economic and financial ambitions were repugnant to Turkey."[85] The vulnerability of Smyrna exposed the weakness of "grandiose schemes of financial and capitulatory control."[86] Archaeologists began to participate more and more in schemes to spirit out antiquities.

Butler ordered Berry to arrange for all material, including the trunks of material from the 1918 Karo expedition to Sardis, to be shipped to him at the MET in New York City. The initial 1918 assessment and agreement to export one trunk of personal items and notes had been in consultation with the Ottoman government. The decision to export more than that was not. Sardis was still officially under an Ottoman permit, despite Greek claims to the contrary. In preparation for export, Berry networked through the pro-Greek Horton and Sardis friendships in Smyrna and beyond. Sardis Society members Robinson and Osborn of the MET endorsed export without permission from Constantinople.[87]

Berry's first export of material from the site on behalf of the Sardis Society took place in 1921. In consultation with Horton and Sterghiades, Butler determined that Greece held legal title to the territory of Sardis, and therefore in December 1921 asked for permission from S. A. Papafrango, Envoy Extraordinary of the Royal Government of Greece, to export materials from Smyrna to New York.[88] Papafrango responded on March 2, 1922:

> I repeatedly wired to competent officers of my country asking that they accept your demand. There are difficulties on account of the archaeological law of my country but I am sure that our officials will examine the question in the most favorable way, because I am in a position to know that the work done by the Americans is extremely favorably looked upon by my country.[89]

That spring a small team, including Berry, two Princeton students (roommates) Lansing "Denny" C. Holden Jr. and Gordon McCormick (Cyrus' son), visited Sardis to collect remaining items that the Sardis expedition intended to ship to the MET and Harvard's Fogg Art Museum; in addition, Denny and Gordon prepared illustrations of the Temple of Artemis, especially the capitals.[90] While these men studied, measured, and illustrated, other teams assessed and began preparations to box and export materials to the United States. Headlines in the *New York Times* before and after the arrival of material in the United States conveyed the unfailing support for Americans in Anatolia.[91]

Crates of material intended for shipment were stored in the warehouse of the Standard Oil Company.[92] Authorities turned a blind eye, owing to pressure and networks of Edward Capps, Horton, Sterghiades, and Cyrus McCormick, whose connections to Standard Oil proved indispensable for arranging logistics. McCormick's relationship with Venizelos and his business connections (via International Harvester) with the Imperial Ottoman Bank (joint venture with British, French and Ottoman capital) also smoothed this along. Members of the Bank prepared the final paperwork for export, described as "instruments and the personal effects of Butler," recalling the original 1918 agreement.

McCormick's role in this process was likely mixed. He saw purpose in export, yet also found great future promise in landscape steeped in history. He waxed eloquently in his April 19, 1922, letter to Butler about Sardis:

> As I stood on the top of the Necropolis I realized the picture that was spread before me. A flood of deep human interest came over me as this panorama of the past moved before my mental vision. The present activities of the excavation; the historic interest of the plains spread out before me like a wonderful carpet; the mystery of the plain of a thousand tombs; the beauty of the surrounding mountain range; the possibility of developing untold agricultural wealth in level land between the mountains; the present political

> difficulties involving two ancient people who apparently cannot reconcile and harmonize their racial antipathies, and above all the keen excitement and thrilling interest of the possibility of discovery in unearthing the records of two thousand years of history—all this combined to make the picture, mental and visual, a thrilling one. I realized your insight and your determination to overcome all obstacles in devoting yourself with such enthusiasm and such success to this inspiring problem. With these and many similar thoughts in mind you can easily understand how gratified I am to have had a glimpse of this strange and interesting country.[93]

On April 10, 1922, the MET had agreed to pay two thousand dollars for shipping of material "upon the understanding that all objects thus brought, which are suitable for exposition in this Museum, shall become the property upon their arrival and acceptance."[94] In total, some fifty-five to sixty crates of material were shipped to the MET. Goode and Yegül argue that Butler intended to return material once the situation stabilized. In confidence, Butler wrote to Osborn, then a trustee at the MET, that repatriation would eventually be necessary or it would jeopardize the future of excavation at Sardis. He underscored that once conditions allowed that half of the finds must be returned to Turkey.[95] Butler confirmed his position with Robinson on April 22, 1922.[96]

Still, the matter appears to have never been fully resolved. The Standard Oil Company of New York, who housed the crates in their warehouse, requested on June 9, 1922, the following: "In view of the fact that an opportunity may arise for the shipment of your antiquities to the U.S.A., we shall be glad to know whether the permission of the Turkish authorities has been obtained in addition to that of the Greek authorities."[97] With this request, Standard Oil made an attempt to remain neutral to protect their interests in Smyrna until jurisdiction was clear.

The deceitfulness involved in this multi-year program that sought not only to pursue export but also to shift the legal protocols resonates in two letters. These show also the precarious position of Butler. The first is from July 8, 1922,

> I, H. C. Butler, Director of the American Society for the Excavation of Sardis, do hereby certify that the contents of the trunk now being shipped by me from Smyrna to New York and addressed to my agent, V. Evertt Macy of 128 Broadway, New York, contains only articles owned and used by me between March 1910 and June 1914 in my home at Sardis and in hired rooms in Smyrna; and that the contents of the trunk are as follows:

Household effects, rugs, carpets, etc.
One rifle, purchased in America.
One shotgun, purchased in America.
Two cameras, purchased in America.
Clothing, used.
A few pieces of antique pottery.

Subscribed and sworn to before me, Maynard B. Barnes, Vice Consul of the United States of America at Smyrna, Asia Minor, this eight day of July 1922.
Official seal
Vice Consul of the United States.

A few days later, on July 11, 1922, Buckler wrote to Butler:

> Were you amazed and flabbergasted and utterly delighted at the sight of thirty chinking golden staters? Surely you clasped the Hodja to your steaming bosom and called upon his God and yours to witness your fine frenzy. . . . What things you will have to show me in the Metropolitan! When I said goodbye to Shear in Smyrna just after the utter collapse and capitulation of the Greek Archaeological Director (he wept salt tears like any crocodile) he was planning in his mind, I think, the removal of the Acropolis. Money from the Museum is now a certainty, of course, and if they are the gentlemen I assume [them] to be, the cheque should be a fat one. May Allah walk with thee, O, Buyuk Celibi! Chabuk bir bashka sefer sen bakajegim, inshallah."[98]

Butler never read this letter, nor did he ever unpack the material. Nor did he oversee the repatriation. He died unexpectedly in Paris in the summer of 1922.

On September 6, 1922, Turkish forces stormed Salihli, a mere 6 kilometers east of Sardis. Fearful of the outcome of Turkish-Greek confrontation in Smyrna, the French, from their battleship in the bay of Smyrna, wrote to Mustafa Kemal to offer Smyrna. Halide Edib Adıvar chronicled Mustafa Kemal's response: "Whose city are they giving to whom?"[99] After torching Salihli in their retreat, Greek forces continued their scorched earth policy as Turkish forces pushed them toward Smyrna. On September 9, Turkish forces entered Smyrna, and in the conflict fires were set. The flames were fanned by high winds. When the smoke cleared days later, most of Smyrna was in ashes, and this event has served as the collective memory of the Turkish War of Independence, now known as the "Asia Minor Catastrophe" (see Introduction and Chapter 3, this volume).

The Treaty of Lausanne was signed on July 23, 1923, and with it the recognition of the sovereign boundaries of the Republic of Turkey. Despite the efforts of the Sardis Society to pressure the US State Department to reinstate Sèvres Article 421 on antiquities, Turkey refused to negotiate. Absent from the final Lausanne agreement were provisions for an international observatory to manage antiquities, archaeological sites, or museums. An ugly legal case ensued between Turkey and the United States regarding the material illicitly taken from Sardis. A statement had been made: Turkey would handle her own affairs.[100]

Discussion

The story of Sardis between the years 1910 and 1922 demonstrates the importance of investigating the scales of impact that archaeologists may have regarding political and cultural sovereignty. This analysis reveals the struggles in which American archaeologists, museums, and their patrons found themselves regarding access to antiquities and new anti-acquisition policies. It also shows the wider participation of archaeologists in diplomacy and international arenas that were integrated into the process of negotiating boundaries at the end of World War I, and again during renegotiations subsequent to the establishment of Turkey.

My research sheds light also on the importance of studies of development during this period in history. Much has been written on the Mandates, the economic undertones of the League of Nations, even the strength of Smyrna as an Ottoman port. Yet, an understanding of how these spheres came together to facilitate Princeton's work at Sardis remains in its infancy. Among the values of studying this narrow window of time in a specific city and her hinterland is demonstration of the inseparable relationships between pedigree and education. Men of "good breeding" had extensive connections to wealth and power. They also embraced notions of growth, progress, and modernization. These benchmarks for "civilization" became *the* pivot points for Greek occupation. Allied tacit approval stemmed from the unwillingness to recognize the fractured relationships driven by greed, and in turn the fear of upsetting beneficial economic relationships. The landscapes of Sardis became a theater of political performance. During the very months, even weeks, when educated men were parading their support for global peace in Paris, tensions tightened and conspiracies conflicted within the British-backed Greek occupation of Smyrna. For the Americans with an interest at Sardis, antiquity was privileged over human life and dignity in the circumvention of legislation to extract antiquities from Sardis and export them from Turkey.

In sum, the cultural dimensions in the wake of World War I and during the Turkish War of Independence form critical components in the history of the Republic of Turkey and international heritage policy from three perspectives. First was the entitlement that stemmed from philhellenism and associated colonial networks; the second reflected the emergence of assistance in the name of revealing history that had been neglected or was yet to be appreciated; and the third was an uneasy realization of territorial loss and thus the failure of the imperial project, even as it was recast under the League of Nations.

CHAPTER 3 | Open Intelligence

Marshall gave me my land.

—FARMER, PROVINCE OF MANISA, western Turkey, 2010

The goods and machinery shown in this exhibition are some of the results of the continuing industrial research in America. They have been selected to show those which can be useful in Turkey to further develop the economy, to increase production and exports, and to build a better life in factories, farms and homes. The exhibition suggests new ways of progress for the future. It expresses a message of American interest in Turkey and the friendship of the American people for the people of Turkey.

—US PRESIDENT JOHN F. KENNEDY'S remarks for the opening ceremony at the 1961 Izmir International Fair

DID FAIRS HELD IN the heart of Izmir's Kültürpark, itself built on the ashes of the Armenian quarter of Smyrna, deliver JFK's promise for "the friendship of the American people for the people of Turkey"? Analysis of the annual fairs at the Kültürpark reveals the tensions between Soviet and American approaches to rural development and their respective foreign policies. At the fairs, participants engaged in, as described by US officials, the practice of "open intelligence."[1] Among the new machines on show at the fairs were those for agriculture, notably Soviet, Czech, French, and, later, US tractors. In these ways, the Kültürpark became a strategic stage of international trade diplomacy and corporate contract culture.

Grafted onto the floor of the Gediz Basin, stretching more than 100 kilometers from the Gediz Delta to the Demirköprü Dam, was a planned countryside—regulators, canals, villages, agricultural fields, and roads. Today the hydrological and irrigation infrastructure represents the ideological entanglements of industrial agriculture and Cold War diplomacy. Ultimately Turkey sided with the United States. Under this relationship, individual sovereignty flourished, owing to the privatization of land and support for neoliberal policies, which put in motion a way of life that has forever altered the face of this landscape and the lives of those residing there. This chapter explores the heritage of Izmir's agricultural trade fairs with the development schemes in the Gediz Valley.

Development Diplomacy

This story begins where Chapter 2 ended, the formation of the League of Nations. Under the League there was considerable focus on cooperation among interest groups, foundations, and governments. Recalling the activities of the Royal Institute of International Affairs (1920), the Institute of International Education (IIE, 1919), and the American Council of Learned Societies (ACLS, 1920), the League pursued cooperation with academia, government, and the private sector through the International Committee on Intellectual Cooperation (ICIC, 1922). At this moment in history, intellectual elites sought to define how the West could and would "improve" the underdeveloped areas of the world. This rhetoric of assistance was carefully deployed to secure international concessions and to pursue strategic economic goals that promised to strengthen US-Turkish relations.[2]

During the interwar period, hydroelectric technologies aimed at generating power, increasing land reclamation, and extending irrigation systems were the hallmarks of broad-scale urban-rural integration and river basin development schemes.[3] These included the massive Dneprostroi Hydroelectric Dam in Russia, its associated water management systems, and the resulting major shifts in agricultural production. At the time it marked a milestone in Soviet-American relations. The unveiling of the first Soviet five-year plan (1928–1932), too, was widely watched by Western leadership. However, in the wake of the 1929 financial collapse, Soviet and American approaches began to diverge. Their respective political ideologies clashed in their quest for development, visions for increasing the industrialization of cities, and the dramatic transformation of the countryside.[4] Turkish architects followed these concepts closely throughout the 1930s; they were published in prominent Turkish publications such as *Mimar, Arkitekt*, and *İnsan*.[5]

These debates were pivotal in shaping social engineering projects, especially as they pertained to how government officials and planners envisioned "civilizing" the "impoverished" countryside while also providing energy for expanding urban centers. In the United States, President Roosevelt launched two signature programs under the New Deal: the 1933 Tennessee Valley Authority Act (TVA) and the 1936 Rural Electrification Act (REA).[6] TVA was "a company clothed with the power of government but possessed of the flexibility and initiative of private enterprise."[7] Its flagship project was Norristown, located in the shadow of the Norris Dam (1933–1936). The project included a substantial social component. Stories in *TIME*, the *New Yorker*, and the *Nation* discussed the social agenda of Norris as the perfect antidote to the Great Depression: community living, leisure parks, and collective spirit.[8] The success of this and other TVA projects drew international attention.

Sir Julian Sorell Huxley, a British citizen, an evolutionary biologist, and the future Director General of the United Nations Educational Scientific and Cultural Organization (UNESCO), admired the art of political persuasion in the TVA approach. Huxley visited TVA, and in 1935 published his *TIMES* essay, "The TVA: A Great American Experiment." In 1942, Huxley and the Chairman of TVA, David E. Lilienthal, shared a number of correspondences, including those that discussed the global expansion of the TVA model. Lilienthal appauled Huxely's admiration of the larger scheme:

> The important news about the TVA these days is the mounting interest in this project and its methods, frequently in a detailed way, manifested by observers from foreign countries. During the past several weeks we have had visitors, official or otherwise, or specific inquiries through foreign government sources from a dozen or so places outside the United States: Sweden, the Union of South Africa, New Zealand, Australia, China, etc. And interestingly enough, the debate about an issue expressed by the various parties to the controversy in terms of "a TVA on the Danube." The fire started when the Vice-President, commenting in an important speech at a Russian friendship meeting, used TVA as a symbol representing the postwar period. This was followed by an attack from the opposite quarter by the President of the National Association of Manufacturers, with a reply and renewed initiative being taken in a speech the other night by Senator George W. Norris. Because you have understood the real point of the TVA so much better and so much longer than most folks, I thought this would interest you.[9]

In his 1943 book, *TVA, Adventure in Planning*, Huxley admired TVA's ability to harness "the democratic principles of persuasion, consent, and participation."[10]

To be sure, the integration of the rural and urban through assistance programs wasn't new. Louis Pauly argued in the mid-1990s, and Fredrick Peterson, Jens-Wilhelm Wessels, and Patricia Clavin have followed, that the template of Western economic assistance formed under the League of Nations was inherently self-serving.[11] The League's committees sought to bolster Western-centered concerns and economic policies. In this way, conference diplomacy of elites (diplomats, economists, engineers, architects, etc.) also came to serve as the League's primary provocateur. International venues often offer places for not only official business, but also informal engagement of state policies and future planning. The specified public goal of ICIC was to foster the international exchange of ideas; yet the underlying agenda was to be sure that these programs satisfied the social and political frameworks of the League's elite and their constituencies.

The ICIC supported international conferences on a variety of topics, including those of the Congrès Internationaux d'Architecture Moderne (CIAM). Established in 1928, CIAM shaped the La Sarraz Declaration of the same year: "Urban planning is the organisation of all functions of collective life in the city and in the countryside." CIAM's best-known member was the influential, charismatic, and visionary Charles-Édouard Jeanneret, known as Le Corbusier. His travels in Latin America, Soviet Russia, the Balkans, Turkey, and America gave him much to consider.[12] The tensions between socialist and capitalist approaches marinated in Le Corbusier's mind and, in so doing, permeated his designs.[13] CIAM's fourth conference, "The Functional City," was planned for Moscow, yet owing to political tensions, it took place on board the S.S. Patris cruising from Marseilles to Athens. The 1933 conference favored modernist agendas of standardization and industrialization in urban planning. The result was CIAM's 1933 "Charter of Athens."

Contrary to their intentions, CIAM's members never completed an ideal model for a city.[14] Many members favored a "green city" concept that followed a collective, communal approach to planning. Le Corbusier, especially, focused on "the 'real needs' of life."[15] The fifth CIAM conference held in Paris in 1937, the same year as the World's Fair "International Exposition of Art and Technology in Modern Life," drew from Le Corbusier's *Radiant Farm* (1933–1934) and *Radiant Village* (1934–1937) projects. Here Le Corbusier considered decentralization and the potential of mechanized agriculture to revive family farms. Careful territorial planning, he argued, would reduce the imbalance between urban and rural lifestyles.[16] His interests included the TVA model, and he visited TVA's headquarters in 1945.[17]

In fact, while both CIAM and TVA promoted designed landscapes for cities, towns, villages, and water management infrastructure, the respective social and political paradigms clashed, one favoring communal investments, the other the promise of the private market and support for the entrepreneur, which the United States viewed as critical components of its new foreign policy. In 1944 Lilienthal's *Democracy on the March* was released. He promoted TVA as a "catalytic agent" and a diplomatic engine that rivaled that of the US State Department.[18] The TVA took great pride in sharing its "American accomplishment" and promoting the global "TVA family."[19]

The success of TVA graced the pages of major news publications: the *New York Times*, the *Wall Street Journal*, and the *New Yorker*. The program was branded "an effective instrument of foreign policy" and "an excellent opportunity for our Nation to demonstrate before the world its ability to harness its resources in basic public works of peaceful intent."[20] David Ekbladh has argued that this policy represented a deliberate strategy to entice planners at home and abroad.[21] The TVA model became the bedrock for many aspects of the Truman Doctrine, the European Recovery Program/US Marshall Plan, and the Point Four program. These initiatives complemented wide-scale US support for the United Nations (1946) and NATO (1949).[22] The resulting tensions propelled Cold War containment polices in river basins. The technocracy of powersheds were bankrolled by the expanding corporatism of water diplomacy.[23]

The World Bank (1944), the International Monetary Fund (1945, IMF), and the United Nations Development Programme (1965, UNDP) all played strategic roles in providing the initial financial means to realize these initiatives. Resulting policies included the US 1954 Agricultural Assistance Act (Food for Peace) and the 1961 Foreign Assistance Act that created the United States Agency for International Development (USAID). Scholars have demonstrated the dubious nature of development structures and egregious gatekeeping; in fact, most studies show that citizens in need often became victims of promises that were rarely, if ever, realized.[24]

With this history in mind, I now turn to one of the earliest programs to follow the TVA model abroad: the Aegean-TVA in the Gediz Valley of western Turkey. Among the goals of the United States was to counter a growing Soviet influence, which was apparent at the Izmir annual fairs and embedded in the design of Kültürpark itself. Both politically and intellectually the ideological vision for Turkey shifted considerably between 1935 and 1955. Sentiment during the 1930s allowed for shades of gray regarding socialist approaches, yet by the early 1950s these perspectives were no longer tenable for the United States. American diplomats wielded significant influence to shift "hearts and minds" toward a capitalist perspective.

Foreigners at the Kültürpark

After 1923, US diplomats and businessmen viewed Izmir as "lost" and dismissed the potential for revival.[25] This was not the Turkish perspective. The city was chosen as the venue for the First Turkish Economic Congress. This sent a firm message to the Turkish public and the international community that Izmir would continue to be a place for business. The Congress helped to imbue the national consciousness with a priority on agriculture and development. The song "Agriculture March," released on March 4, 1923, spoke to the heart of Turkey. As shown in Figure 3.1, the lyrics indicate that villages and farms would sustain the nation.

The only foreign ambassadors to participate officially in Izmir's 1923 Economic Congress were those from Azerbaijan and Soviet Russia. Because the conference took place at a space provided by the Hills Brothers, Inc., of New York, honored guests included the US High Commissioner, the American Vice Consul, the senior US naval officer in port, and a handful of American businessmen. Less than a month after the

> Confidently we plow, sow, and harvest
> Ensured that each bit of the yield belongs to the nation,
> We responded to Head Farmer Atatürk's call,
> Gathered on the agriculture front, for the battle with the soil.
> We are the foundation, the roots of the nation.
> The real owners of the homeland, the masters, the villagers.
> This earth, this ancestry, makes us human.
> Working earnestly, using the newest tools,
> As a Turk, we shall be an example to the neighboring world,
> An honest, vigorous mind with calloused hands, and a heart at ease,
> In our homeland, we shall build wealth, order.
> National sovereignty shall tear down all obstacles,
> Let the villages enjoy plenty, comfort and joy,
> That is us, the invincible Turkish soul.
> We are the foundation, the roots of the nation.
> The real owners of the homeland, the masters, the villagers.[26]

FIGURE 3.1 English Translation of Turkish Lyrics to *Agriculture March*.

Congress, Atatürk unveiled "Nine Principles" (*Dokuz Umde*) that rallied a nationalistic sentiment and celebrated the electoral base of farmers and merchants.[27] The Congress put in motion efforts toward land reforms, including privatization, technological improvements in water management, new approaches to agriculture, connectivity of the nation through roads and railways, and a commitment to the social life of "peasant classes" and, ultimately, as Kaçar writes, the framework for what Atatürk sought to create: "a new Turkish citizen."[28]

The presence of Russian (formal) and US (informal) representatives at the 1923 Congress foreshadowed future interests and entanglements. The Soviet connections in Turkey strengthened during the 1920s. This friendship was in part a result of Russian assistance to Kemalist resistance against the British, French, Italians, and Greeks during the final hours of the Turkish War of Independence.[29] Prime Minister Ismet Inönü, Foreign Minister Tevıf Rüştü, and thirty-two other members of the Turkish delegation visited Soviet Russia in 1931 Members of the delegation toured cultural areas, attended social events, and visited industrial sites.[30] A key part of the trip was to discuss the Dnieper water management and hydroelectric infrastructure, the urban renewal programs in Moscow, countrywide industrialization, and the five-year plan model.

In turn, Inönü invited a group of Soviet consultants to Ankara. The Turkish-Soviet relations were featured in Lenfilm's 1934 release of the movie *Ankara—Heart of Turkey* that reaffirmed this decade-long alliance. Thus, when the Russian delegation came to Turkey to consult on the first five-year Turkish industrial plan, also unveiled in 1934, it was within this spirit of collaboration between the two nations.[31] The same year the Soviets awarded assistance of 8 million USD to Turkey.[32]

The Turkish-Russian relationship was also cultivated by Russian participation at the Izmir International Fairs. They attended the first in 1927 and continued their participation throughout the 1930s, along with a strong showing from Egypt and Greece (Figure 3.2). Russian agricultural machinery was a key feature at the Soviet Pavilions. These displays were far superior to those presented by the few American companies who attended.[33] The Soviets also leveraged the political cache from the massive Dneiper project. In fact, this design was infused with US expertise from an American commission appointed as consultants, notably experts from the International General Electric Company.[34] And, in fact, it was very much this work as well as major dam and irrigation projects in the United States that structured the TVA.

FIGURE 3.2 Greek Pavilion, 1935. Izmir International Fair. Courtesy of the Ahmet Piriştina City Archives and Museum, Izmir.

During this same period of time, Turkey explored and adapted European models of rural land reform (and reclamation) and urban plans.[35] The 1924 Beaux Arts plan for Izmir was designed by brothers Rene and Raymond Danger (with contributions from Henri Prost, also known for his plans for Istanbul), and focused on integrating residential and business districts around open boulevards, tree-lined promenades, and public parks. On the ashes of Alsancak and Basmane scorched by the 1922 fire, Rene and Raymond proposed that Izmir bury this past under a Kültürpark, a space that would give a solid nod to the future of industry and modernism.[36] Originally planned for 36 acres, it was expanded to 360 acres (Figure 3.3). The inspiration for expansion came via Suad Yurdkoru, chairman of the Izmir Soccer Association. He had led a 1933 sports diplomacy program to the Soviet Union, where he visited Moscow's Gorky Park, which had opened in 1928.[37] Once appointed deputy major of Izmir, Yurdkoru persuaded Izmir's mayor, Behçet Uz, to expand Izmir's Kültürpark along the lines of Gorky. Experts from Moscow came to Izmir to assist with design.[38]

The unveiling of the Kültürpark by Prime Minister Ismet Inönü on September 1, 1936, gave Izmir a tremendous international platform. The job of the organizers was to attract foreign investors, whether it was through foreign assistance or market-driven profits. They were successful.

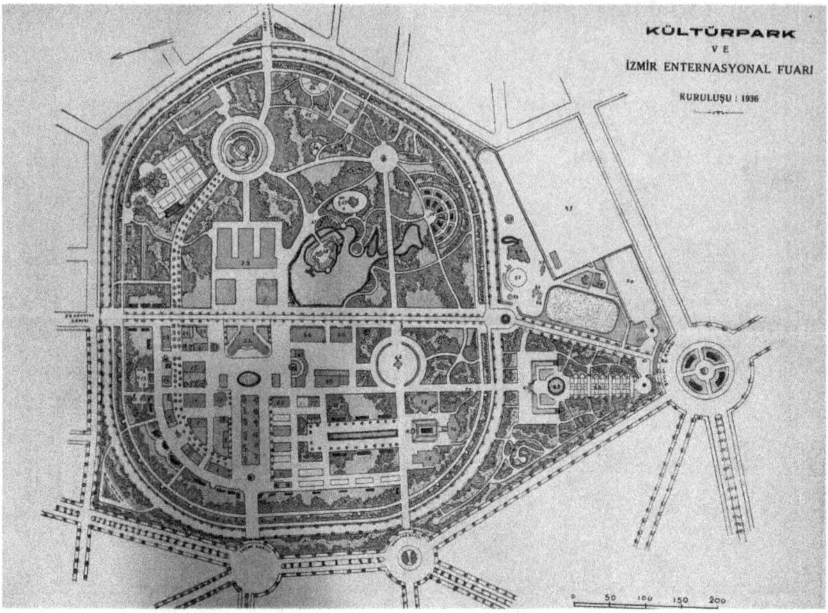

FIGURE 3.3 Plan of Izmir's Kültürpark, 1936. Courtesy of the Ahmet Piriştina City Archives and Museum, Izmir.

The park was a central meeting hub, not only for businesses but also for prominent social gatherings. Americans began to realize their oversight in not making a solid effort to be well represented at the annual fairs—a place where Germans, Greeks, Egyptians, and Russians were hob-knobbing with their Turkish counterparts (see Figures 3.4 and 3.5).[39]

Running parallel to the growing prominence of the Kültürpark were plans for the redesign and expansion of Izmir. In 1938 Le Corbusier was contracted to prepare a new plan. The pillars of CIAM were closely followed in Le Corbusier's illustrations and report: *habiter* (dwell), *travailler* (work), *cultivar le corps et l'esprit* (cultivate body and mind), and *circuler* (circulate). A series of city plans, one shown in Figure 3.6, illustrated access routes (111-4), residential zones (112-1), work zones (112-2), and culture-nature zones (112-3)—all within the garden-city paradigm.[40]

Le Corbusier took into consideration Izmir's regional and national contexts, which is unsurprising given his focus on radiant farms and villages and his visit to TVA. He completed five separate plans: Physical Geography (101-1); National and International Connections (101-4); Urban Izmir: Distribution and Agglomerations (102-1); The Centers of Production (102-2); and Centres of Culture (102-3). Figure 3.7 illustrates

FIGURE 3.4 Pavilion of I. G. Farben Industry, Frankfurt. 1938. Izmir International Fair. Courtesy of the Ahmet Piriştina City Archives and Museum, Izmir.

FIGURE 3.5 Soviet Pavilion, 1938. Izmir International Fair. Courtesy of the Ahmet Piriştina City Archives and Museum, Izmir.

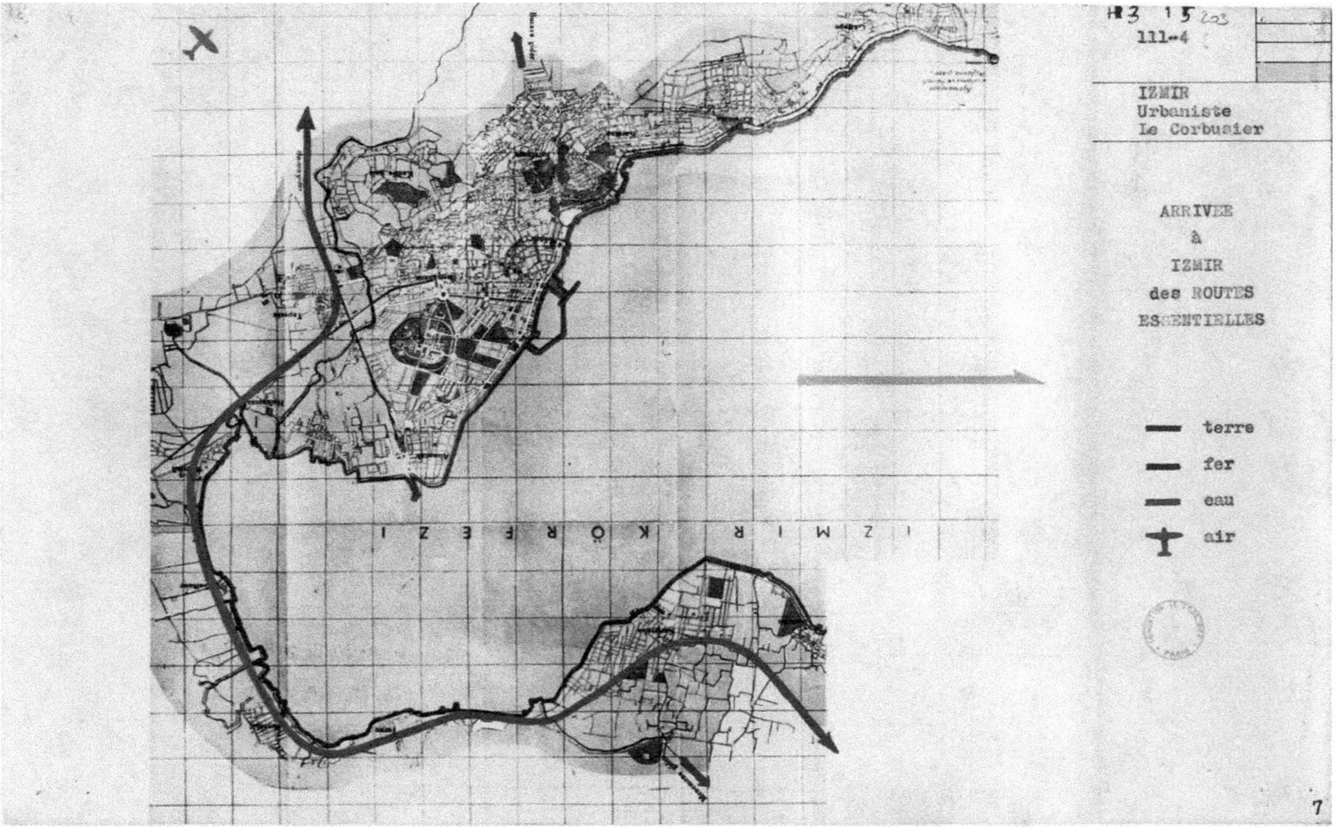

FIGURE 3.6 Le Corbusier's Izmir Plan, 1949. H3-15-203-001. Courtesy of the Le Corbusier Foundation.

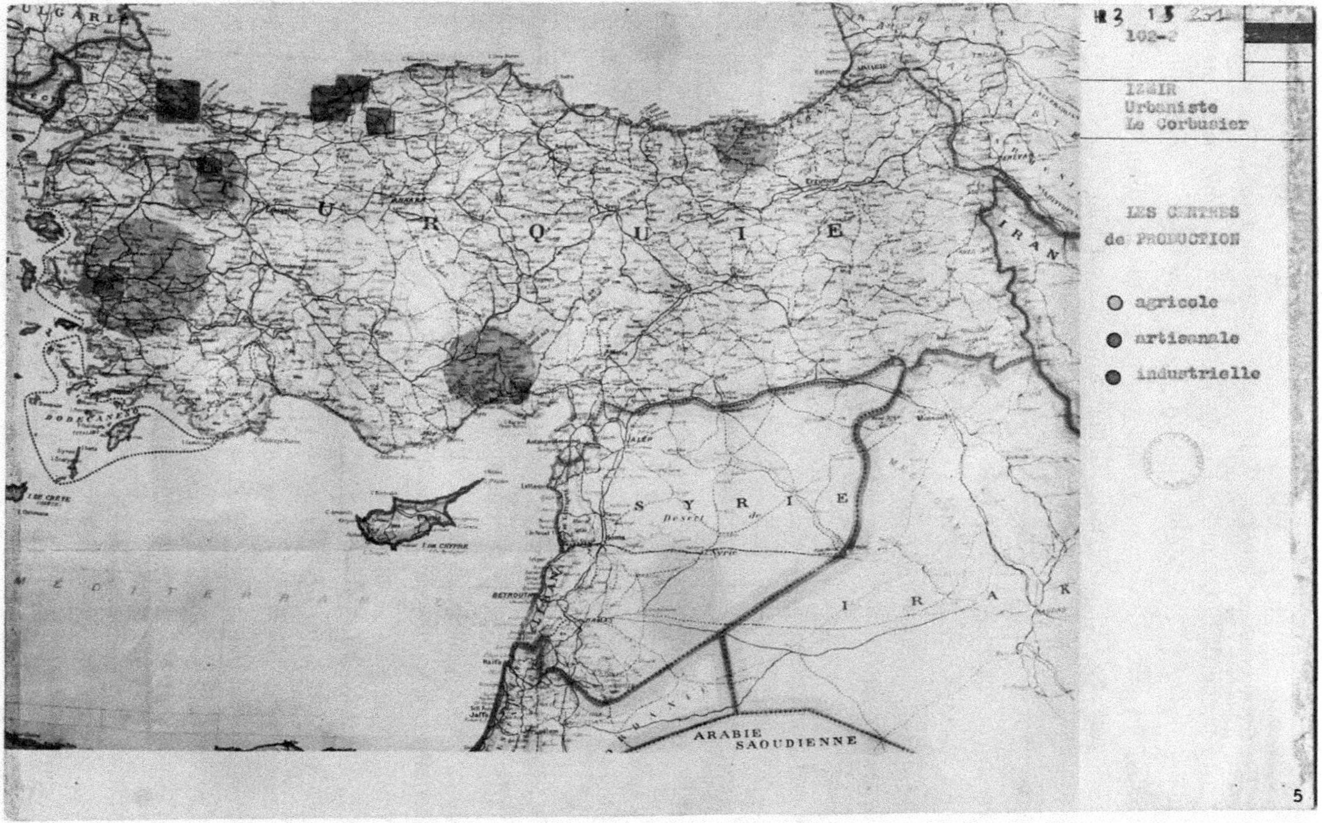

FIGURE 3.7 Le Corbusier's 102-2 Agricultural, Artisan, and Industrial schematic map for Turkey, 1949. H3-15-201-001. Courtesy of the Le Corbusier Foundation.

the plan for the Centers of Production (102-2). The goal was to demonstrate that without the countryside, the city could not thrive and vice versa. Planners spoke of a mutual interdependence between the rural and the urban.

Yet, Le Corbusier's idealism included radical new territory. As Bilsel writes, Le Corbusier advocated for "the abolishment of private ownership of property in urban land in order 'to permit human movement in all directions.' "[41] This utopian ideology may have been fitting for the late 1920s and early 1930s, but World War II and the fracturing of Turkish-Soviet relationships and the rise of anti-communism[42] made Le Corbusier's proposal presented in 1949 untenable. Consideration of communal property contrasted with neoliberal tendencies and new laws for rural areas, which corresponded to the 1926 Civil Code No. 743, implementation of the 1934 the Cadastre and Land Registry Law, and the 1936 Land Registry and Cadastre Directorates. On June 11, 1945, the Turkish Grand National Assembly passed the "Law Providing Land for Farmers." These various components of legislation vested land with those who had farmed it for more than twenty years, or to whom experts deemed the rightful owner, further outlined in the June 4, 1947, "Regulation on the Distribution of Land to Farmers."[43] These ideas were also embedded in Turkish policies for rural development.

Turkish-American Alliance at TVA

Between 1946 and 1969, US leadership sought to influence and capitalize on Turkish policies of urban renewal and rural development.[44] At the same time, TVA expanded its diplomatic efforts in a "Training and Educational Relations Branch" charged with connecting the Turkish Embassy in Washington and US academic institutions (e.g., Harvard University, the University of Michigan, and Yale University). On May 22, 1947, US Congress put forth Public Law 80-75: Assistance to Greece and Turkey. This was the Truman Doctrine and allowed the United States to finance projects in countries not eligible under the European Recovery Program (i.e., the US Marshall Plan). It transformed US-Turkish relations.[45] The International Bank for Reconstruction and Development (IBRD) also endorsed this new US policy, reporting that "[i]n order to lighten the financial load arising from the containment of Russian expansion in the Mediterranean and Middle East, the United States Government in 1947 gave a $100 million grant-in-aid to the Turkish Government."[46] Results came quickly.

On July 10, 1947, officials signed the "Agreement between the Government of the United States and the Government of Turkey Regarding a Highway Program." The US War Department contracted with the Gary Slag Corporation in Chicago. The program included trans-country highways and rural, village "farm-to-market" routes, which was modeled after the Missouri program under US Public Works Association grants. The corporation claimed to build three miles of roads per day.[47]

A top secret 1947 memo discussed additional aid to Turkey to raise the "standard of living of the Turkish peasant and workman" in order to stay off "infiltration of Communist ideas."[48] The US government did eventually extend Marshall Plan funding to Turkey, a decision recognized publically on radio stations and privately within diplomatic circles.[49] A strategic goal rested on overhauling the rural landscape and expanding the family farm. This was a tall order: 80 percent of Turkey's population in 1945 lived in small villages. Turkey circulated a document entitled "About the Marshall Plan" to businesses and municipalities. It included instructions for how to use new technologies for farming, where to attend trainings (in Turkey and the United States), a list of reputable repair shops for machinery, and payment options: loans through Ziraat Bankası (State Agricultural Bank), cash, or special agreements with specific companies.[50]

Connections at the Izmir International Fair helped secure these types of arrangements. American and European companies with ambitions to garner the market on the booming business of mechanized agriculture attended (Figure 3.8). The 1947 poster (and a 1-lira stamp) of a dove seen weaving a multi-nation banner through the Kültürpark's towers (Figure 3.9) reflects the hope for peace and mutual understanding. This theme is made all the more clear with further context: the towers flank the Lausanne Gate, and thus symbolize the idealism of international peace following the end of World War I.

In addition to the hope of contributing to world peace, in 1947 the Izmir Fair became a member of the Global Association of the Exhibition Industry. In these ways, the opening of the 1947 Fair (Figure 3.10) marked a re-entry of Izmir onto the international stage. The entrepreneurial spirit was matched by the jockeying of divergent political perspectives in Turkish foreign policies, particularly Russian-Turkish and US-Turkish relationships.

In 1948 the US and IBRD financed mapping programs in Turkey.[51] If private property could be registered with proper boundaries, then the owner could participate in assistance programs. These initiatives complemented the road program financed under the Truman Doctrine. Implementation

FIGURE 3.8 Steyr Tractor Pavilion. In foreground is 180 model from 1947. Izmir International Fair. Courtesy of the Ahmet Piriştina City Archives and Museum, Izmir.

of private property boundaries, however, met with ideological challenges. Increasingly, policies functioned on the principle of eminent domain through forced expropriation in the name of the state. The Turkish Ministry of Agriculture was given authority to enforce expropriation, and the State Agricultural Bank handled the financial transactions and awarded "special credits" for farmsteads, equipment, and seed. This complemented funding from the Marshall Plan.[52]

International Harvester, which first entered Turkish markets earlier in the century (Chapter 3, this volume), offered loan packages for tractors and other farm equipment. The company ran small- and large-scale advertisements featuring images of tractors and "farm machines" in the *Cumhuriyet* newspaper. As of 1948, "only a few thousand hectares of State-owned land ha[d] actually been distributed to farmers, and no private estates ha[d] been expropriated."[53] Two years later, on September 18, 1950, the US Foreign service outlined why a focus on the Turkish farm was a prudent US initiative: "65% of village farm families have sufficient land and capital to make a fair living."[54] Farmers would be most successful, the reports argued, if they had access to large fields, new infrastructure for irrigation, and mechanized equipment to implement "power farming." Sales

FIGURE 3.9 Izmir International Fair, 1947 Poster. Courtesy of the Ahmet Piriştina City Archives and Museum, Izmir.

of tractors took off, as shown in Table 3.1. The favorable bottom line was also in large part due to the low cost of human labor.

US researchers continued to gather key information that was then translated by the US Office of Foreign Agriculture Relations. This included data pertaining to individual versus cooperative ownership of machinery (e.g., tractors, plows, diggers, sprayers, selectors, etc.), manufacturers (e.g., John Deere, International Harvester, Caterpillar, Massey Harris, etc.), and lists of dealers.[55] There was a good showing of US companies at the 1950 Fair (Figure 3.11).

The US news, too, presented the case for American involvement in Turkish agriculture: "[T]he people are modern-minded and eager to

FIGURE 3.10 Opening of the Izmir International Fair, 1947. Courtesy of the Ahmet Piriştina City Archives and Museum, Izmir.

TABLE 3.1 Regions in Turkey with over 200 tractors by 1950

REGION	TRACTORS IN 1948	TRACTORS IN 1950	PERCENT INCREASE
Aydin	134	265	98
Konya	54	250	363
Manisa	75	201	168
Seyhan	650	790	22
Tekirdağ	62	202	226

learn, not frozen in ancient traditional habits. Atatürk brought about such drastic changes in the 15 years of his rule—from fez to hat is a greater psychological jump than from ox to tractor—that the masses of this semi-Oriental country are surprised by nothing," wrote Thompson in the *Daily Boston Globe*.[56] Between 1948 and 1950 roughly 6,500 tractors (valued at $14,857,000) had been imported as part of the US Economic Cooperation Administration program.[57] Subsidies, credits, and loans for tractors and other equipment, especially packages of equipment, were directly tied to projected crop yields.[58] The success of the US assistance programs depended on the willingness of American businesses to invest, and both US President Dwight D. Eisenhower and Turkish Prime Minister Adnan Menderes gave their endorsements. The impact of Menderes's policies

FIGURE 3.11 Izmir International Fair, 1950 poster. Courtesy of the Ahmet Piriştina City Archives and Museum, Izmir.

for harmonizing unregulated development in urban centers has been the subject of much research, especially for Istanbul.[59] It also was the topic of diplomatic cables to assess the relative impact of foreign assistance[60] and large-scale investment, particularly regarding the storage and transfer of oil[61] and water management for hydroelectric projects.

Fuat Köprülü, Foreign Minister under Menderes, supported the vision of the Democrat Party: "a policy of extreme liberalism in economic as well as political fields."[62] On May 21, 1950, Köprülü requested a meeting with Mr. Winthrop Aldrich, Chairman of the Board of Directors of the Chase National Bank, a significant investor in the Middle East at the time. Increased foreign investment and free enterprise meant tough choices for Menderes' government, especially if Turkey was to attract American capital for development projects. Köprülü struggled with expropriation, which he felt was "incompatible with the notion of lawfulness and equity."[63] This concern paralleled matters before the Turkish judicial system related to immovable property and enforcement of regulations allocating lands to farmers.[64]

On September 17, 1951, the US Embassy in Ankara sent a memo with the subject line "TRACTORS ON FARMS—TURKEY" to the US Department of State. It detailed the approximate numbers of tractors on family farms throughout Turkey as well as the number on order.[65] Follow-up reports to the Department of State through 1952 and 1953 confirmed sustained demand for new equipment.[66] By 1954, the equipment program had spurred policy changes for crops based on global markets (figs, raisins, wheat, sesame, and olives), and in turn changes were made in Turkey to protocols for mapping, deeds, purchases, and expropriation.[67] Simultaneously, the US Embassy worked to position key US companies such that they could secure contracts for machinery, equipment, planning, and construction.[68]

The success of the US road project in Turkey led Mr. Kasım. Gülek, Turkish Minister of Public Works, to seek collaboration for drainage of swamps and the construction of hydroelectric and irrigation plans along the "Authority model."[69] This interest had been fermenting for some time. Through the US Office of International Information and Cultural Affairs at the Department of State, TVA officers set up visits for Turkish engineers to tour Norris Dam, Chattanooga Dam, and Widows Creek Steam Plant. During these sessions, the Turkish delegation discussed technical expertise, installations, combustion engines, institutional structure, project organizational charts, and financing.[70]

In turn, the Turkish government hired US consultants to evaluate river valleys, including John L. Savage, an American hydro-engineer with dam-design experience (and a former member of the US Bureau of Reclamation). In 1949, in preparation for his trip to Turkey, Savage visited TVA.[71] In fact, by the early 1950s, visits to TVA became the new normal for Turkish diplomats and engineering students.[72] Visitors included Mr. H. C. Gökçen, the first General Director of the Turkish Departments of Electrical Research, and later the Director of Eti Bank, which financed large dams in Turkey. Gökçen had been part of the UN mission to the Middle East to establish an "Arab TVA." The former chairman of TVA, Gordon Clapp, led the group.[73] Recall that J. Huxley had met Clapp during his visit to TVA. Gökçen, Clapp, and Cabir Selek, the Executive Director for Turkey to the IBRD in Washington, D.C., also held discussions.[74] Negotiations for financial arrangements for dams in Turkey went forward. When Mr. Emin Yumer replaced Gökçen, he also visited TVA.[75] Turkish Minister of Agriculture, Esat Budakoğlu, too, made a trip to Tennessee in the fall of 1956.[76]

The Turkish presence at TVA throughout the 1950s and a robust engine of US consultants in Turkey led to re-evaluation of US participation at the Izmir Fair. In fact, one of future US Ambassador Gilbert Robinson's first major assignments with the Eisenhower administration included a trip to the Fair. He was told that the "United States [was] conspicuous by its absence at international trade fairs. [They] want[ed] to do something about it."[77] The US Embassy wrote in 1950, "International Fairs offer excellent opportunities for the dissemination of general information about this country, its geography, culture, folk-ways, etc., as distinct from commerce."[78] Robinson described a special commission that was set up between the US Information Agency and the US Department of State to gather information about trade shows. Eisenhower also directed funds for the Office of International Trade Fairs, established in January of 1955.[79]

This same year Robinson was asked to attend the Izmir International Fair. He initially declined because of his lack of experience. His superiors, however, felt that his experience in public relations and his youth (he was 26) would be great assets. Robinson described the Fair as "electric" and an ideal place for "open intelligence."

> We [the USA] put on a major exhibit there. The Russians were there with a big exhibit because they had just had to come down to the Black Sea with lots and lots of lumber. The Israelis were there, as were the Romanians. It was a miniature World's Fair. It was very interesting for me as a young man. The Russians used to try to find out what we were doing, so they would wine and dine the Israelis and then later on we would wine and dine the Israelis and find out what the Russians were asking.[80]

The Fair became a strategic theater for Cold War performance. The US Embassy in Ankara determined that the venue offered great "psychological" aspects of diplomacy. In order to compete with the Soviet exhibits, the US Embassy recommended three components: (a) film and art for "maximum psychological impact"; (b) official US government and US private exhibits; and (c) a professional design team to stage and coordinate the exhibit (and to outshine the Soviets).[81] The showing of US films *Hoover Dam, Factory Worker Turns Farmer*, and *Valley of the Tennessee* were all woven into the new fabric of selling US aid abroad.[82]

A decade on, 1960 marked an auspicious time for water diplomacy and by extension plans for agricultural expansion throughout Turkey. The US government made a commitment to maintain a prominent, sustained presence at the 1960 Izmir Fair (Figure 3.12). According to the report from the US Embassy for that year, the US pavilion drew the largest crowd, with the Russian pavilion the next favorable. The US theme was the port

FIGURE 3.12 Sketch of the USA pavilion and photo of US President Eisenhower, 1960. Courtesy of the Izmir International Fair. Ahmet Piriştina City Archives and Museum, Izmir.

FIGURE 3.13 Izmir International Fair, 1961 Chamber of Commerce. Courtesy of the Ahmet Piriştina City Archives and Museum, Izmir.

of Philadelphia, chosen to highlight its "cradle of liberty." But, during the 1960 fair, Turkish officials quietly questioned US commitment and credibility; in response, the US government made overtures to attend future fairs.[83] Just one year later in 1961 (Figure 3.13), substantial funding for the Fair came from US Public Law 480 (PL480), the source of a parallel stream of funding that was allocated non-coincidentally to nearby archaeological projects at Sardis (see Chapter 5, this volume).

While the influence of TVA and Le Corbusier reflected the waxing and waning of ideological perspectives in Izmir, from the planning of the city itself to the annual Izmir International Fairs, there was also another component to the US presence in western Turkey. Le Corbusier's plans emphasized the importance of the countryside to the city, and the TVA model, too, stressed the importance of this relationship. In this context, I now turn to one of the earliest TVA-inspired projects to be implemented in Turkey: the hydroelectric Demirköprü Dam in the Adala Canyon, the Lake Marmara reservoir, and the irrigation systems of the Gediz Valley.

Marshall's Gift to the Gediz

In the 1930s, most rural territory in Turkey had not yet been mapped and registered in a format that would harmonize with foreign development standards. In addition, municipal and government sectors needed to be organized accordingly. In 1914 the General Directorate of Public Works had been formed, followed in 1925 by the establishment of a Water Science Board and Water Works Directorates with offices in Bursa, Adana, Ankara, Edirne, and Izmir.[84] Throughout the 1920s efforts were made to produce 1/200,000 scale maps, with some cities having 1/25,000 scale maps prepared (including Manisa).[85] After this point, a number of needs coincided: the basic need for water in urban areas; the need for more electricity for growth in industrial areas; and the need to combat the public-health threat from malaria, as shown in Figure 3.14. Common consensus was that hydroelectric dams and infrastructure for controlled irrigation could do much to solve these concerns.

The construction of the Çubuk Dam in Ankara was under way by 1930 and was completed in 1936. At 25 meters high and with a total capacity of 1,200 m^3, it was tiny compared to the massive Dnieper Dam in Russia (1927–1932) and the colossal Hoover Dam (1931–1926) in the United States. The strong Turkish-Russian relationship of the early 1930s spurred discussions for a hydroelectric dam and irrigation project for the Gediz Valley and Izmir. It would dovetail with plans for the region: market expansion of agricultural products and machinery, land reclamation, controlled irrigation, potable water, and reliable electricity.

FIGURE 3.14 Image of sicknesses crippling the Republic, including malaria. Courtesy of the Ahmet Piriştina City Archives and Museum, Izmir.

While the Gediz River had been rerouted through its delta in the nineteenth century, serious work for a hydroelectric dam and permanent reclamation of wetlands had been untenable given the costs and required expertise. In the middle Gediz, annual floods inundated agricultural zones, and every five to seven years catastrophic floodwaters created havoc for rail lines, roads, bridges, farms, villages, and towns. Control of wetlands was given even greater priority with the August 2, 1930, malaria outbreak in Izmir and nearby areas. The following years, 1931–1932, were dry and contributed further to these fears. In 1933 plans were made to dredge a canal from the course of the Gediz in Adala toward Lake Marmara as part of a new reservoir and irrigation system.[86]

Tenders for the project were released in Spring 1930.[87] The project gained international interest, especially in the United States, Switzerland, and Russia.[88] By 1935 infrastructure for the expansion of Lake Marmara into a reservoir had been started, and in 1937 an electrical study for a hydroelectric dam was completed. Plans for a valley-wide irrigation system were discussed in 1938, which also led to negotiations for aerial imagery, even the potential for photogrammetry (at the time a cutting-edge technology).[89]

On February 20, 1942, the Gediz River overflowed her banks, inundating the plains of Manisa and Menemen, including roads and rails. Reports of extensive damage to crops and villages pervaded Turkish newspapers. On May 10, 1944, the Republican Peoples Party held a conference to discuss continuing widespread damage. The winter of 1948 brought more extensive periods of rain and massive pools of water. On June 1, 1948, the Turkish Ministry of Public Works estimated that Turkey would require 165 million Turkish Lira for the construction of extensive water-control systems to manage the Seyhan, Sarkaya, and Gediz rivers. These projects would also provide energy to major cities and infrastructure for reclamation of rural territory for agricultural purposes. On January 3, 1949, *Cumhuriyet* published a call for tender for the Gediz hydroelectric facilities.

On May 25, 1949, students under the direction of Professor Ilhan Çetin from the Geological Institute in Ankara visited Manisa for the newly launched aerial survey of the Gediz Valley. During the same month a Yale University group toured the area.[90] This type of people-to-people dialogue defined US strategy and involvement in the Gediz. It followed the slow, deliberate, and pervasive approach to development diplomacy marked by collaboration among "experts" that Huxley extolled, and that the League's ICIC, too, had sought.

In 1949 the Department of Electrical Research under the Turkish Ministry of Public Works hired the Knappen, Tippetts, Abbett Engineering Company of New York (KTA). This was among their first major foreign contract in the region. In subsequent years the company, then Tippetts, Abbett, McCarthy, and Stratton (TAMS), were contracted for plans and construction of the Seyhan project in Adana and the Kemer Dam in the Büyük Menderes.[91] The company then moved south. Between 1952 and 1957, TAMS had contracts for the Tigris-Euphrates Valley and the Basra region in Iraq, which competed with Soviet contracts for dams in Syria and Egypt.[92]

KTA's report, *Gediz Basin Study*, was submitted to the Turkish government in 1950.[93] It provides a lens into the heart of US water diplomacy.

The report claimed that infrastructure would improve "stability of employment . . . the promotion of education, and the provision of public services"; even military efforts would improve. Expansion in agriculture and electricity "would increase available private capital, foreign purchasing power and the supply of essential raw materials needed for initiating industrialization, and . . . provide the power required for substantial industrial expansion."[94] KTA gathered data from the relevant industrial firms poised to benefit from the project,[95] yet no interviews were done with communities immediately affected (i.e., those living in the middle Gediz Valley). Impacts on their lives were determined by scientific analysis of the terrain, and from these results areas likely to benefit from land reclamation were determined, including viable terrain for agricultural fields and new towns.

KTA assessed the topography and hydrology of the region to determine the optimal locations for irrigation canals and regulators. The report explained that "Priority was given to lands promising the greatest and earliest monetary returns to assure a self-liquidating project and to facilitate financing. . . . In general, suitable areas were only found in the main river valley below Adala and in the lower portions of the valleys of the major tributaries in the same region. . . ." (KTA, Ch. V, 15). As to eventual water administration, KTA recommended the "Authority" model (i.e., the TVA).

Management at a regional level would cultivate leadership and encourage the "active participation of the people of the region." A Gediz Valley Development Board was to include "seven outstanding leaders of the community," such as "representatives of farmers, bankers, industrialists, major communities and the public." Board members had to be "the most distinguished and public-spirited members of the community" who were "able, civic-conscious local men familiar with the regional problems and conditions." The United States believed this type of structure would enhance its credibility over the long term and provide the social ammunition and ideologies necessary to thwart Soviet influence.

The initial infrastructure for Izmir's electricity, supplied by a Belgian firm, was spotty at best; electricity for small industrial plants was cut a half-hour before sunset and restored after 11:00 p.m.[96] Most telling was that 67 percent of rural areas and most villages lacked access to any electricity at all. For KTA these rural communities were not communities in need of assistance, they were future customers—at least 130,000 of them. Further analysis promoted a "redoubling model." KTA estimated that returns would come in as few as two years, and as long as seven.[97] KTA described these projects as making communities "industry-conscious." Diplomatic practice sought to encourage technological expertise in order to increase the Valley's "potential foreign exchange value."[98] In sum, the private sector had the potential to profit handsomely once the project was completed.

Regardless of the promised stability from the project, plans for the Gediz hit trobled waters very quickly. The World Bank had commissioned an analysis of Turkey's economy. The result was the 1951 Barker Report, which encouraged technological developments that would promote agricultural productivity. On November 17, 1953, World Bank representatives visited the region with the explicit purpose of assessing the viability of a loan to finance the hydroelectric dam at Demirköprü. This same year, the World Bank questioned Turkey's credit, and ultimately refused to consider the Demirköprü project. Owing to these unfavorable reports, in March 1954 the World Bank was ousted from Turkey.[99] The Turkish government absorbed the majority of the costs for Demirköprü. In partnership with the newly formed State Hydraulic Works (Devlet Su Işleri, DSI) in 1954, Turkey began to expropriate lands that would be developed as part of the water management system. It was for this reason that mapping in the middle Gediz was fast-tracked.

The significance of the project was reported in 1955 in the *New York Times*: "western Turkey's first large multipurpose hydroelectric project, patterned after the Tennessee Valley Authority." The formal project tag was "Turkey's Aegean T.V.A." In addition to containment policies, deeper histories were also used to validate US attention to the project, and especially to Izmir. "The Eastern Question" and the "loss of Smyrna" (see Chapter 2, this volume) crept into the *Times* essay: "to restore the economic primacy to this area that was enjoyed under the ancient Greeks" by removing "flood and silting danger on the Gediz River delta north of Izmir (Smyrna)." The project would assist in US efforts to help "close the economic gap."[100] Not only would there be potential for kickback profits to American companies (e.g., sale of farm machinery at the Izmir International Fair), but the project put in motion greater opportunities to leverage US political currency over the long term.

In 1954, the TVA was the impetus for awarding the first Eisenhower Fellowship in Turkey for travel to the United States. The recipient was Süleyman Demirel, then 31 years old: "Mr. Demirel has, as his principal area of interest, national resources development and is particularly interested in learning some of the activities of the TVA."[101] It was Demirel's second visit; the first was to the US Bureau of Reclamation in 1949, just after he graduated from Istanbul Technical University. This specific sponsorship as an Eisenhower Fellow is particularly interesting because the fellowship program was the brainchild of the US Central Intelligence Agency (CIA).[102] The timing of these various visits parallels US interest in Turkey that extended far beyond agriculture. This same year saw the realization of NATO. Turkey was a strategic partner, becoming a member in 1952. These training programs were part of significant US commitments to containment, and contributed to the work of the Joint American Military Mission for Aid to Turkey (JAMMAT), established in 1947 in Ankara.[103]

After his time at TVA, Demirel returned to Turkey to become the Head of the Department of Dams (1954) and shortly thereafter the Director General of the State Hydraulic Works (1955–1960). He was known as the "King of Dams."[104] Demirel took a keen interest in Demirköprü.[105] His interest was helped by the politics embedded in the project. In June 1950, Member of Parliament Fevzi Lütfi Karaosmanoğlu had been charged with implementation of the Marshall Plan; he later became Minister of Interior in 1951, giving him even greater authority.

Olivier Bouquet argues that major Ottoman elite families and specific individuals often became vital figures in Republic-era agendas.[106] The choice of Lütfi as MP ensured a strategic voice in the province of Manisa. Lütfi knew the region well, and he had trained as an agricultural engineer (in Istanbul). His family stood to lose significant land holdings in the national program to vest possessors and occupiers with ownership. Yet, they could also gain lands, especially areas that would be drained and reclaimed as part of the Demirköprü project, lands close to the Bağırgan Çiftlik, his family's nineteenth-century estate just east of Bin Tepe in the Gediz Valley (see Chapter 6, this volume).

The Bağırgan estate provided a central hub for planning and networking. Agricultural reports (subsidies, crop yields, and more), proceedings from Parliament, drafts of speeches, personal letters between Demirel and the family, maps, and even a copy of KTA's 1950 *Gediz Basin Study* remain in the estate's private collection today. In keeping with the KTA's suggested "Authority model" to appoint vested members of the community to overview the project, Demirel vested Lütfi with oversight of those responsible for the Gediz Basin water project.

The modifications to the environment were extreme. They required moving massive amounts of earth and blasting on a scale that had never been seen in the region. Loud, dusty construction zones contrasted the pristine design plans KTA had provided as well as the sleek hydroelectric components arriving from France, including bright orange turbines made by Neyrpic (later Alstom PH). As designs turned into reality, the landscape modifications were awe-inspiring for many. Project leadership and support staff lived nearby in a company compound:[107] manicured gardens, soccer fields, outdoor and indoor theaters, a tennis court, an open-air pool, a billiards room, gazebos overlooking the massive Demirköprü reservoir, and a tiered system of living quarters spaced along suburb-like streets.

Reports of the first detonation of dynamite in the Adala Canyon resounded quickly and far. People watched with apprehension and amazement as the machines from abroad rolled into town (e.g., bulldozers, backhoes, and more). The process was not without apprehension and concerns about security and safety. The *New York Times* reported that "The [Turkish] Government alerted 100 villages for possible evacuation today after heavy rains burst an auxiliary wall at the Demirkorpu Dam. . . . No loss of life was reported, but damage was said to be heavy."[108]

The imprint of the dam, the spillway, and the central electric station required changes at the juncture point where waters of the Gediz would be channeled into the valley's irrigation systems and the Lake Marmara reservoir. Adala became the canvas for a ready-made blueprint to accommodate this 1950s mega-project. A massive regulator was laid over the Gediz, and channels for the left and right banks of the irrigation systems were integrated into the old town as well as areas slated for future development. Below the regulator and among the canals, areas were designated for small parks and cafes.

To control the Gediz meant a fundamental shift in life for those who owned land in the former flood plain, now to be crisscrossed by new irrigation channels and sub-channels. Many of these changes would capture pubic imagination, garner support for the government, and place far greater agency in the individual, now made possible by land ownership and agricultural production.

As of the spring of 1954, however, lands in the Marmara Lake Basin and the Gediz Valley had yet to be officially mapped or registered. The formal process of expropriation required transfer of deeds, and thus formal titles needed to be issued. Farmers who could claim title based on legal possessor and occupier stipulations were given proper deeds. If expropriation was necessary, they were then paid for the land, often mere hours after being issued title. In contrast, those in areas not subject to expropriation were granted and maintained official title to their land. Today, some farmers remark with pride, "Marshall gave me my land," a tribute to the catalyst of the irrigation project, the US Marshall Plan.

Once complete the monumental imprint of the irrigation canals and regulators defined the face of the Gediz Valley: a series of parallel and interconnecting ribs embossed on the valley floor. Prior to the canal system, there were no formal roads, nor bridges. Transit routes that were once extremely challenging over wetland areas became more reliable, especially those that crossed the Gediz, now made possible via two new routes leading north from Salihli and Ahmetli.

On May 18, 1960 Prime Minister Menderes opened the Demirköprü Dam. Thousands came to cheer the accomplishment as well as a particular vision for the future of Turkey. Less than two weeks later, on May 27, 1960, the entire project and region fell under a dark shadow. A military coup unseated Menderes. Charged with violating the constitution, he was sentenced to death and executed on September 17, 1961, despite pleas from various world leaders, including US President John F. Kennedy. Newspapers suggested that dubious financial arrangements regarding the construction of Demirköprü may have played a significant part in gathering support for the coup.[109]

Demirel's position in this sphere of politics is not to be underestimated. After the coup, he worked as an independent consultant and taught at the Middle East Technical University (which had also been established in part with US Marshall Plan funding).[110] He served seven terms as Prime Minister of Turkey, the first beginning in 1965 and the final ending in 1993, when he became the ninth President of Turkey (1993–2000). He opened the Izmir International Fair in 1967 (Figure 3.15).

Discussion

The actions of foreign consultants in the territory from Izmir to Adala between 1923 and 1960 balanced Turkish policies with Soviet and later US interests. At the core of foreign activities were ideological underpinnings of economic development and the construction of civic life. The reconstruction plans for Izmir waffled between the *Beaux Arts* paradigm of the Danger brothers and the CIAM vision of Le Corbusier. Both emphasized the Kültürpark and the International Fair as the critical civic and economic catalyst of Izmir, yet Le Corbusier's vision for urban-rural connections pushed this aspect considerably. Competing notions of private property and economic assistance through the development of the city's hinterland were informed by discussions with the Soviets and Americans regarding the viability of five-year plans and variations in how Turkish partners conceptualized how life was to be lived, whether in cities or on farms.

FIGURE 3.15 Prime Minister Süleyman Demirel, 1967. Izmir International Fair. Courtesy of the Ahmet Piriştina City Archives and Museum, Izmir.

Major shifts in geopolitical arcs after World War II shifted Turkey's alliance toward American development diplomacy and the TVA model. The emphasis on accessibility to the hinterland and water diplomacy to secure hydroelectric power for the urban core as well as to shape the mechanized and newly designed territories (roads, bridges, canals, land reclamation, etc.) in the Gediz followed US policies that sought to bolster US economic interests. Izmir's Kültürpark and Adala's Demirköprü Dam—and all associated infrastructure that was grafted onto the face of the Gediz Basin between these two landmarks—encapsulate the ideological tensions at play concerning economic and agricultural foreign assistance to Turkey.

The networks embedded in water diplomacy paid handsome social and political dividends. As I've shown, those vested in interwar policies, organizations, and new technologies held leadership positions at UNESCO.[111] The TVA became a critical node. Huxley met Clapp through TVA, and later Huxley's position at UNESCO gave him the power to appoint Clapp to lead the Middle East mission on water diplomacy. The other major nexus point was the connection between Turkish engineers at TVA and those who joined Clapp's mission, who in turn were introduced to key members of the World Bank and the IBRD.

The most critical axis centered around Süleyman Demirel. His time at TVA gave him considerable prestige on the global stage. In addition, he respected and leveraged legacy Ottoman-era networks. The role of Fevzi Lütfi Karaosmanoğlu demonstrates the strategic political climate at the time, and the significant position of the Gediz Valley from the perspective of Ankara. As a member of the pre-1923 Karaosmanoğlu dynasty, Lütfi would have known well the major businessmen of Izmir, those who would have stood to profit from expansion in electricity as well as increases in agricultural produce. He was well connected with the primary banks and had previously overseen the programs for loans and subsidies meant to promote the sale of agricultural machinery. His family estate in the heart of the region under development made him a particularly valuable partner.

In Chapter 5 I turn to the restoration programs at Sardis, which, similar to the water diplomacy initiatives, were handsomely funded at the start. Yet, unlike the water and agricultural programs, the cultural projects were abandoned by the US Department of State by the mid-1960s. In Chapter 6, I return to the Marmara Lake Basin and the complex histories of the industrial infrastructure described in this chapter, the pursuit of farming, and its impact on the archaeological and natural heritage of the region.

CHAPTER 4 | Diplomatic Deliverables

I consider it to be in the interests of the United States to assist in rescuing these [Abu Simbel] historical remains of a former civilization from destruction—and to join the international effort. . . . By thus contributing to the preservation of past civilizations, we will strengthen and enrich our own.

—US PRESIDENT JOHN F. KENNEDY'S public speech, April 7, 1961

As you know, the Roman gymnasium (so-called Bedesten) is preserved to a very considerable degree. It faces on the wonderful and impressive Roman marble avenue. Both, from the point of view of location and from the point of view of the setting of antiquities, it would make an excellent museum, if suitably restored. I wonder what you think of this possibility.

—DIRECTOR OF SARDIS AND PROFESSOR AT HARVARD UNIVERSITY, GEORGE M. A. HANFMANN, to General Directorate of Archaeology in Turkey, June 30, 1962

RUNNING PARALLEL TO THE water management and agricultural development projects that came to define the Cold War industrial heritage grafted onto the floor of the Gediz Basin in western Turkey, as discussed in Chapter 3, were plans for research and restoration at the nearby archaeological site of Sardis. Yet, unlike the success of US foreign policy to support rural farming communities through Aegean TVA programs, especially efforts to back individual sovereignty through privatized landownership, attention to archaeological research and preservation was not a sustained US government priority. This despite the prominence of Sardis, the critical economic importance of Izmir for US containment policies, and the international rhetoric for US support of archaeology and preservation as revealed through the UNESCO salvage campaign in Egypt. The notorious Nubian campaign showed temples and monuments inundated by the rising waters from the Soviet-built High Aswan Dam. The relocation of the Abu Simbel complex set up an expectation for proactive US policy toward preservation and archaeology as a critical component of Cold War diplomacy.

Unbeknown to the US public and the global archaeological community, however, the United States had no intention of committing resources to heritage preservation at the level of Abu Simbel for other areas of the world. In response to the expanding budget of the Nubian program, and fear of future restoration projects, on September 4, 1963, the US Foreign Relations Committee cut expenditures for archaeological fieldwork and preservation from future State Department budgets, including those for the archaeological work at Sardis in western Turkey under the PL 480 (Agricultural Assistance Act). Apparently there is no official record of these deliberations.[1] My knowledge of this decision and its relationship to the Nubian program has been achievable only from analysis of correspondence among archaeologists and members of the Smithsonian Institution and US Department of State archived by the Archaeological Institute of America (AIA), American Schools of Oriental Research (ASOR), the Archaeological Exploration of Sardis, and Cornell and Harvard universities.

Unlike the early twentieth-century heritage games of jurisdiction, archaeology in the mid-twentieth century operated under a different set of practices. Foreign projects were no longer attempting to smuggle out antiquities. Rather, they had dedicated research plans and sought to work with the Turkish Ministry on tourism initiatives to create archaeological parks. Restoration of the Marble Court-Bath Gymnasium and Synagogue complex at Sardis provided the blueprint for the restoration of major monuments at the archaeological sites of Ephesus (i.e., the Library of Celsus), Pergamon (i.e., the Temple of Trajan), and Aphrodisias (i.e., the Temple of Aphrodite). In this chapter, I explore how archaeologists at Sardis realized this monumental project in the face of waning US commitments to archaeology in Turkey, even as US Congress and others rushed to protect the monuments of Nubia.

Decade of Development

Abu Simbel has become legendary in preservation circles as a global turning point in heritage studies. Many argue that it was this monument and this moment in time that provided the catalyst for the 1964 International Congress of Architects and Technicians of Historic Monuments. The outcome was the 1964 Venice Charter and the establishment of the International Council on Monuments and Sites (ICOMOS). The momentum provided further support for the realization of the 1972 Convention Concerning the Protection of the World Cultural and Natural Heritage.[2] This is a story of archaeology and development in Egypt jockeying along a strategic geopolitical frontier during a critical period of Arab-Israeli and Egyptian-Soviet relationships. Negotiations for the Suez Canal were on the table. This vital transshipment zone for oil and rising Arab nationalism threatened post-WWII structures of governance that were favorable to Western powers. Efforts to combat these perceived threats were launched through US Army Corps of Engineers in-country projects (military airports, roads, etc.), which supported the United Nations "Decade of Development."[3] While scholars continue to reveal new perspectives on the underpinnings of Abu Simbel, less documented are areas of the world and contemporary projects that were compromised by the massive undertaking in Egypt. This was the case at Sardis.

Three archaeologists were involved in the Sardis case: George M. A. Hanfmann, Jotham Johnson, and A. Henry Detweiler. Hanfmann was born in St. Petersburg, Russia, in 1911. He immigrated to the United States via Berlin in 1940. During World War II he served in London with the US Office of War Information. He returned to the United States after the war and became curator at Harvard University's Fogg Art Museum and in 1956 Professor of Art History. He became the Director of the Archaeological Exploration of Sardis in 1958. At the time, Jotham Johnson was a Professor at New York University, Director of Aphrodisias, and President of the AIA. In his role with the AIA, he advocated for fieldwork as an invaluable part of US diplomacy. The AIA's partner institution for the Near East, eastern Mediterranean, and northern Africa is ASOR. From 1955 to 1966 Detweiler served as ASOR's President. He was also Professor of Architecture at Cornell, and in 1956 was appointed Associate Dean of the College of Architecture. He then became the Associate Director of Sardis.

Detweiler's early work in the Middle East and Mediterranean included extensive time in Jordan at Jerash, where he participated in the survey and rehabilitation of monumental architecture.[4] He later prepared the *Manual of Archaeological Surveying*, published in 1948 by ASOR with a subsidy from the American Council for Learned Societies.[5] Correspondences from colleagues at Yale University, the University of Michigan, and the University of Pennsylvania, among others, indicate the tremendous contribution of the *Manual* to the field of archaeology. This work was important for Detweiler's professional reputation as an architectural archaeologist.[6] This pedigree earned him a place on the US Advisory Committee on the Salvage of Abu Simbel (see Table 4.1).

TABLE 4.1 Members of the U.S. Advisory Committee to the Salvage of Abu Simbel. A. H. Detweiler Archives, Cornell University

DETWEILER ARCHIVES, CORNELL UNIVERSITY. US ADVISORY COMMITTEE ON THE SALVAGE OF ABU SIMBEL	
Mr. Harold Blum, Jr. Briar Hill Stone Company Clermont, Ohio	Dr. Edmundo Lassalle Vice Chairman for Public Relations Nubian Committee 320 Park Avenue New York 22, New York
Mr. Edward B. Burwell Geological Engineer Upperville, Virginia	Mr. Robert Nesbitt Corps of Engineers Department of Defense Washington 25, D.C
Dr. A. Henry Detweiler Associate Dean College of Architecture Cornell University Ithaca, New York	Mr. William Stevenson Smith Curator, Dept. of Egyptian Art Museum of Fine Arts Boston 14, Massachusetts
Mr. John Dimick Archaeologists 2029 Connecticut Avenue, N.M. Washington 8, D.C	Mr. Edward E. White Spencer, White & Prentis, Inc. Contractors and Engineers 10 East 40th Street New York 14, New York
Mr. Francis Donaldson Chief Engineer Mason & Hanger – Silas Mason Co. 300 Fifth Avenue New York, New York	Dr. John A. Wilson Director, Oriental Institute The University of Chicago Chicago 37, Illinois
Mr. Wendell E. Johnson Chief, Engineering Division Civil Works Office of the Chief of Engineers United States Army Washington 25, D.C	Mr. Leon Zach Landscape Architect Corps of Engineers Department of Defense Washington 25, D.C
Battle Ass for Cult Aff [scribbled in pencil]	

On August 27 and October 31, 1968, former Ambassador to the United Arab Republic (1964–1967) Lucius (Luke) Battle was interviewed by Larry J. Hackman for the US Ambassadors Oral History project.[7] Prior to Battle's post in Egypt, Battle was the US executive secretary to the US National Commission to UNESCO (1961–1962) and was charged with US policy regarding Nubian monuments. He served as the chairman of the 1962 UNESCO General Conference held in Paris to discuss Abu Simbel. He continued as the primary US representative on matters regarding Abu Simbel as Assistant Secretary of State for Education and Cultural Affairs (1962–1964). In his 1968 interview, Battle described these years as "a kind of merger of my hobbies and my career in a nice sort of way, and I had a great many friends in the educational world and the cultural world and the foundation world." Battle's colleague Max McCullough, Special Assistant to the Office of the Assistant Secretary of the Bureau of Educational and Cultural Affairs at the Department of State, assisted in his efforts with UNESCO.

During this same period of time, Detweiler, Hanfmann, and Johnson corresponded with members of the Smithsonian Institution who oversaw the funds allocated for US archaeological projects abroad, including pots of money for both Egypt and Turkey. In 1964 Mr. Sidney Dillon Ripley was appointed as the Secretary of International Activities and worked closely with Mr. William Warner (consultant to the Secretary for International Activities, 1964; Special Assistant to the Secretary for International Activities, 1965; Director of the Office of International Activities, 1966; Assistant Secretary for Public Service, 1967–1974). New applications of steel in restoration, such as Albert Caquot's "The Temples of Abu Simbel: Draft Proposal for Moving Them in Entirety," impressed committee members and convinced them to pursue the project.[8] The rising price tag and bureaucracy of Abu Simbel gave many in US Congress pause regarding the high costs involved with deliverables and archaeology.

Gifting Temples

The Nubian campaign was not the first to move toward integrating archaeological research, restoration, and development deliverables. There had been gradual movement toward this ideal for over a century.[9] Among the many hurdles post-WWII was funding for research-driven fieldwork, particularly given that archaeology could not offer the quick turn-around economic deliverables typical of agricultural or extraction projects. In fact, even though tourism at archaeological sites was a primary goal of the US Marshall Plan in Greece, private individuals and foundations paid the bulk of these bills.[10] By the early 1960s, US Congress had devised several avenues for the US Department of State to provide support for archaeological and preservation work abroad. Through Public Law 480 section K (the Agricultural Trade Development and Assistance Act of 1954, also known as Food for Peace), excess currency could be allocated for training in archaeology.[11] In collaboration with the US Department of State, the Smithsonian Institution managed these funds. Detweiler and Hanfmann secured PL 480 funds for Sardis, as did Johnson for Aphrodisias. Sardis also received money from the 1961 Foreign Assistance Act and the 1961 Mutual Exchange Act (Fulbright–Hays Act). In addition, the National Science Foundation (1950) and the National Endowment for the Humanities (1965) supported archaeological research at Sardis, with NEH providing critical funds in the late 1960s.

Support from NSF and NEH stressed the pursuit of knowledge as part of scientific and humanistic research embedded in the Academy. In contrast, the 1961 Mutual Exchange Act dovetailed closer with the models of PL 480 and later USAID. These programs preferred to support opportunities for in-country trainings. For archaeological projects, the goal was to communicate best practices through technical expertise in the restoration of objects and monuments. In this way, archaeology offered intangible skill-based deliverables. This was especially true after policies of *partage* were no longer viable. By the 1960s most national patrimony laws prohibited this type of system. Exceptions were made, however, and Abu Simbel was one of them.

In exchange for support of the Nubian campaign, countries were to be gifted temples from Nubia. The Temple of Dendur at the Metropolitan Museum of Art in New York was Egypt's diplomatic gift to the United States (Figure 4.1). The ability to secure excess currency funds held in Egyptian banks (and losing value given the runaway inflation) for a US enterprise was amenable to US Congress and the US Senate. In sum, US support for the Nubian Campaign was a result of poor planning in other spheres; to allocate funds targeted for agriculture to archaeology was one way to preserve dignity and to bolster the reputation of the United States in preservation.

In this light, US support of Abu Simbel was not the moral commitment to public service or preservation that JFK claimed to it to be. Rather, the project was akin to a private-sector contract with clear, tangible (and valuable) deliverables. Battle courted US Congress and the American people to propel a massive project that served to establish a US footprint in Soviet territory, secure Nubian temples for Western nations, and, perhaps most

FIGURE 4.1 Temple of Dendur. Given to the United States by Egypt in 1965, awarded to The Metropolitan Museum of Art in 1967, and installed in The Sackler Wing in 1978. Courtesy of the Metropolitan Museum of Art.

importantly, to serve the interests of the Kennedys. Jackie's love of antiquity was not insignificant in sealing the deal, and Battle knew it. Debates over the universal goals of archaeology and preservation were secondary for the US government in Egypt.

US Archaeology in Soviet Territory

Here I turn to political geography of Egypt in the late 1950s. Tensions were high and negotiations fierce for the contract to build the Aswan High Dam. Egyptian President Gamal Abdel Nasser Hussein was a pivotal figure. Nasser's positive relationship with the Soviets threatened US interests in the region. The Aswan High Dam followed a history of hydroelectric dams at a time when the Tennessee Valley Authority (TVA) model was a well-established brand. Unsurprisingly, during the planning phases for the dam, experts from TVA provided critical support to Egyptian colleagues regarding agricultural policies and technological planning for hydroelectric power and associated irrigation.[12] As in Turkey's Gediz Valley in the 1950s (see Chapter 3, this volume), changes were made to Egyptian agrarian land laws, including policies for private deeds and expropriation. The Aswan High Dam took on a multi-pronged purpose: improving health by battling the breeding grounds of malaria; increasing food production by expanding agricultural territory and productivity; and furthering economic development by, following TVA models, providing opportunities for farmers to become self-supporting and, thus, also self-liquidating.

The Aswan project was planned to be a collaborative effort with Egypt, the United States, and Britain. The World Bank, too, was at the table. Talks collapsed[13] and in 1959 the Soviets got the contract.[14] The United States persisted, however, in staking claim to their (ideological) footprint in Egypt. In a speech on the thirtieth anniversary of the TVA, May 18, 1963, President Kennedy commended TVA's role in supporting the family farm and the private sector: "In short, the work of TVA will never be done until the work of our country is done. There will always be new places for us to go, for, in the minds of men the world over, the initials TVA stand for progress, and the people of this area welcome progress. And it stands for cooperation between public and private enterprise. . . ."[15] TVA's legacy spread to areas of South East Europe, Turkey, the Middle East, Latin America, and South East Asia. These landscapes became Cold War theaters (Chapter 3, this volume).

On September 23, 1963, Bryant M. Wedge, Director of the Institute for the Study of National Behavior, Inc., in Princeton, New Jersey, lamented the "unfair charges" of 'neo-imperialism' embedded in US policies abroad (including those of TVA) levied by the Communist World.[16] He supported the tenets outlined by Robert Rossow in his essay "The Professionalism of the New Diplomacy," published in *World Politics*: "One of the most important and delicate tasks in professional diplomatic training is to break the student's crust of ethnocentrism without damaging the roots of loyalty and preference for the mother culture."[17] In the sphere of archaeology, this meant exercising technological expertise to justify cultural boots on the ground in contested territories.

The commitment to save heritage was bolstered by calls for the United States to patch up their relationships with Egypt as well as deals for diplomatic gifts. John A. Wilson, Executive Secretary of the US National Committee for the Preservation of Nubian Monuments, wrote on February 8, 1961, of the situation:

> When thereafter the Soviet Union stepped into the picture and offered financial aid for the new High Dam, the relative prestige of the United States suffered another blow. In a direct way, it is impossible to recapture the former esteem and good will. However, generous participation in saving materials which are threatened by the Dam will help greatly in restoring a measure of reputation. Conversely, a failure to participate generously would enforce the local feeling that the United States is indifferent to the welfare of the people along the Nile. This is a cultural matter, not a political matter. By turning to the world organization, the Governments of the U.A.R. and of the Sudan have said that these archaeological treasures belong to the world at large, not to their governments alone. The response by the scholars and scholarly institutions of the world has been wide and immediate, free from national jealousies or competitions. From the scientific side, cooperation and mutual good will have set the spirit. . . . American museums, colleges, and committees have recognized that success in the Nubian program could be rewarding in the return of Egyptian antiquities or even of Egyptian temples, to enrich our knowledge of the past.[18]

Sketches and photographs from early travelers helped to galvanize support for the initiative, such as those shown in Figures 4.2 and 4.3.

Discussion of this topic had begun years earlier. As President of ASOR, Detweiler was consulted in the late 1950s. In 1960 UNESCO's journal *Museum* published the assessment of the "International Campaign to Save the Monuments of Nubia." Christiane Desroches-Noblecourt captured audiences with her title: "The Greatest Open Air Museum in the World in Danger." In October 1961 the *Courier: A Window Open on the World*, published by UNESCO, ran a glossy article with maps and photographs. Entitled "NOW OR NEVER: Abu Simbel," the appeal was firm: the member nations were expected to support the initiative where "Art & Science conspired with Nature." From the archaeological camps and "floating laboratories on the Nile," the region was presented as a united front to save antiquity from the technological marvel of modern hydrological science. Little was mentioned of the massive disruption to people's lives.[19] To be sure, the work put Americans and Europeans on Egyptian soil in the thick of Soviet action.[20] In fact, this was a primary point for committing to the project: "The psychological-political argument carries more weight than the economic one. The Russians are building the dam as strictly an economic enterprise. By saving the temples, the United States could show they care about the spiritual side—and realize the importance of saving the cultural patrimony of Egypt."[21]

FIGURE 4.2 Temple of Dendur. Watercolor and gouache on off-white wove paper. Frederick Arthur Bridgman, 1874. Courtesy of the Metropolitan Museum of Art, 2000.597. Rogers Fund, 2000.

FIGURE 4.3 The grotto temple of Abu Simbel, seen from the Nile. Getty Images No. 463953063.

UNESCO's active interest under the guise of global good and progress represented only part of the picture.[22] Self-proclaimed guardianship of world history gave UNESCO an international voice. Good "friendship" and support to tackle preservation was rewarded with temples. The negotiations took place in September 1963 and were included in annexes of the report of the fifth session of the executive committee for the UNESCO International Campaign to Save the Monuments of Nubia, held at the UNESCO headquarters in Paris, October 7–9, 1963.

The international rally—and commitment—for the project, however, came primarily from Kennedy's April 7, 1961, speech:

> Pursuant to Section 501(c) of the Mutual Security Act of 1954, as amended, I transmit herewith my recommendations for participation by the United States in the international campaign initiated by UNESCO to preserve the ancient temples and other monuments in the Nile Valley which are now threatened with inundation as a result of the construction of the Aswan High Dam. . . I consider it to be in the interests of the United States to assist in rescuing these historical remains of a former civilization from destruction – and to join the international effort. . . By thus contributing to the preservation of past civilizations, we will strengthen and enrich our own. Letter to the President of the Senate and to the Speaker of the House Concerning Preservation of Ancient Monuments in the Nile Valley.

Kennedy made heritage a global concern about the legacies of past civilizations. We could not understand ourselves without the past, and in this case, the past of a civilization as complex as the Egyptians was the business of the Western (white) world.[23]

Luke Battle was the turnkey figure for implementation of the Abu Simbel project because he'd determined from where and how to secure a lot of the money, and he knew it would only be possible if the US President supported it. "It was the interest of the Kennedys, really, that gave the impetus for the project. This is one of the very few projects in the cultural field in which there's been a kind of broad international support. I suppose it's the only." Battle recalled that Mrs. Kennedy "followed what we did and participated. The President also. He would occasionally call me—and this was fairly rare before the Kennedy era, for the President to call a lowly assistant secretary."[24] According to Battle, Abu Simbel was the only cultural project that had electrified such broad international support, and he'd found a way for the United States to make the commitment. The 1954 Mutual Security Act, as amended, allowed for foreign currency loan repayments to be used for in-country projects, and thus the United States "could contribute some of the vast holdings of pounds that really had no purpose, they could not be used outside of Egypt."[25]

Scholarship has given the impression that funding came easily for the "Monuments for Peace."[26] Securing the deal was far more tenuous. To release PL 480 funds in Egypt, Battle needed US Congress to approve the plans and resources.[27] He proceeded with caution:

> [W]e had a request for a twelve million dollar appropriation, and while it was clearly stated in this legislation, in this request, that it would go for the purchase of Egyptian pounds, it was still identified as a dollar appropriation. Unless you read the fine print you never knew what it was all about. When I got to the Congress, in the first case, the House Appropriations Committee, John Rooney [John James Rooney], who was the Chairman of the Subcommittee, turned it down.
>
> And then I appealed it to the Senate Appropriations Committee, and Allen Ellender [Allen Joseph Ellender], oddly enough, not a great lover of culture, saw a technical gimmick that bothered him. He said that we were sitting in Egypt at that time with a large fund growing out of the Cowley [Harold D. Cowley] appropriation for the use of private American business in the country and that after any of those currencies had been held for three years they could be used without resort to the United States Congress because they were available for Cowley loans and no limitation existed after three years' holding. And so I picked up on that suggestion saying I thought this was a very distinguished and very statesman-like suggestion, and I assumed that he applied this provision to Abu Simbel and was willing to see the Department of State spend money from the Cowley funds and that we could, therefore, use these funds without an appropriation, and I would hope that such a directive would be made in the report of the Committee of their willingness to have this procedure take place.
>
> I went back to John Rooney and told John Rooney that this is what had happened. He said he wasn't enthusiastic, but he wasn't going to object. So this ended up going through on a kind of fluke. And I had the feeling, he told me a couple of times, Rooney did, that he was well aware of the interest of the President and Mrs. Kennedy in this. In fact, my recollection is that President Kennedy spoke with Rooney at one point, either he did or someone in the White House did, and I think this is what led Rooney to accept this without objection, and this is how it went through. . . .[28]

The scale of the request was the first and the last of its kind in US history for a single heritage project. The monuments to be moved, as shown in

FIGURE 4.4 Abu Simbel, Transport Ramses Kopf, Getty Images No. 542393753.

Figures 4.4 and 4.5, were truly colossal. The process was an engineering marvel and at a cost never before realized for a heritage project.

Battle saw the dangerous precedent that the project would set; even the Russians agreed: "[The Russians] had, well, I would say almost a contempt for UNESCO, which I did not have. I was frequently irritated with it, but they objected strongly to the tendency toward proliferation of activity, and they felt that the United States was quite right in trying to contain it." Battle feared that other countries with monuments under threat would expect the United States to write future checks:

> Moreover, we were getting into a situation that later became a very key one, and that was whether the recipient nations, having one vote as did the donor nations, could raise the level of the budget with the major contributors opposing that level. If this had been permitted we would have, I think, started

FIGURE 4.5 Modern machinery is used in salvaging the Abu Simbel Temple as part of the Aswan Dam Project. December 28, 1964. Getty Images No 3267957.

> a kind of practice that would have caused very serious repercussions, here, on the willingness of the US Congress to go on appropriating money in a situation over which they had no control. It bordered on the ridiculous from our point of view. . . .[29]

This context gives us further background for understanding why the US Foreign Relations Committee made the decision that they did on September 4, 1963, regarding federal support for archaeology. Because this decision was made prior to the final vote for appropriations for Abu Simbel (November 16, 1963), it had to be kept quiet. President Kennedy's 1961 speech had been portrayed as the US commitment to global heritage as part of good foreign policy. Behind the scenes, however, Battle had to strike a careful balance to secure funds for Abu Simbel, yet promise nothing else. This placed all foreign US projects that had received government funds for archaeology in grave jeopardy, including Sardis, among the projects that had been granted stimulus under PL 480 for archaeology *prior to* Abu Simbel.

Battle leveraged the need to spend US dollars that had been converted to foreign currencies that had become increasingly less powerful. That is, section K of PL 480, was not set up necessarily to do good things; it was set up to avoid the negative fallout from runaway inflation. The American Anthropological Association's promotion of the Smithsonian's program couched the initiative in more diplomatic terms: "to provide local currencies for those expenditures that can most advantageously be met overseas."[30] The real reason, however, was to spend what amounted to the dregs of poorly invested US dollars.

Shadow over Sardis

Similar to Egypt, the Cold War dictated US water and agricultural interests in western Turkey. And, similar to Abu Simbel, Sardis offered an opportunity for multinational collaboration, supposedly a pillar of the Mutual Exchange Act of 1961. But, unlike Abu Simbel, Sardis offered no booty for export. That is, the mission focused on research for knowledge (in the name of science) and opportunities for Turkish-American collaboration. By 1964, however, two things began to plague funding for Sardis: no promise of *partage*; and no clear, short-term (two- to three-year) multiplier-effect deliverable that had a benefit to US companies. For these two reasons, the United States turned away from archaeology in Turkey.

This policy, however, went against the grain of what Sardis had planned, with US support in 1961. From the start, Hanfmann had believed that holistic excavation, strategic restoration, and comprehensive diachronic surveys could complement holistic regional planning and that participation by US academic institutions in this process was not only good archaeological practice but also progressive diplomacy. The US Embassy in Ankara, too, agreed, as did the Turkish Ministry of National Education.

This was an auspicious time for Turkey. Prime Minister Adnan Menderes had been overthrown in a military coup on May 27, 1960, and subsequently executed (hanged) on September 17, 1961. Under Menderes, US-Turkish relationships had seen promise, especially in the early years, and the integration between urban centers and rural towns and villages, too, had prospered, especially through the infrastructure projects. In addition, Menderes had ushered through military provisions. These included the placement of nuclear-tipped Jupiter missiles at the airfield in Çilği just north of Izmir. The deal was struck with Menderes in 1959, but by the time they were installed in 1961, they were largely obsolete. Even so, the symbolic nature was valued by the Turkish government (despite American pleas to forgo the operation).[31]

The Cuban missile crisis further spooked Turkey. On October 23, 1962, Prime Minister Ismet Inönü wrote to JFK:

> Whatever may be the repercussions and developments, you may rest assured Mr. President that Turkey will remain faithful to her close ties and to the obligations of her alliance with the United States of America. . . . I have also the pleasure to inform you that instructions have already been sent to the Heads of the Turkish Missions both to the United Nations and to the North Atlantic Council to ask them to give their fullest support to the United States Government.[32]

In return, commitment from the US to Turkey continued. On the public stage, this was reflected in US participation in the Izmir International Fair (see Chapter 3, this volume), and yet much of the support was for the military, a point R. W. Komer expressed concern over: "Turkish political outlook is quite uncertain and we see trouble ahead. Moreover, our very sizable aid to Turkey (about $170 million MAP and $120 million economic in FY 63) seems overweighted [*sic*] on the military side at a time when our top problem is to develop a viable state."[33] It was against this backdrop that the Sardis Expedition proceeded, and while it may not have had the high-level security and immediacy of these other events, it did offer a neutral, and softer, image for US-Turkish relationships.

Even so, the cultural dimension of US support in Turkey did not receive the fanfare that it did in Egypt. On one hand, there wasn't an impeding disaster on the scale of massive inundation akin to that of Nubia, but on the other, there was opportunity for more concrete and reflective planning, especially in the countryside of the Izmir. This would give the United States an additional feather in their hat in a region that had not only received commitment under the US Marshall Plan for design of water management, but also a region intimately tied to the urban hub of Izmir, and thus the international fairs, the strategic deep port, and the military airfield at Çiğli.

Funding for the initial years of work at Sardis (1958–1962) came from Hanfmann's and Detweiler's home institutions, Harvard and Cornell universities (each contributing ca. $10,000 per year), and the Turkish Ministry of National Education.[34] Behind the scenes, Detweiler pursued federal funding, including efforts to secure PL 480, Section K funds. Detweiler outlined the plan in his May 30, 1960, correspondence to Cornell President Deane W. Malott:

> We are starting negotiations with the State Department for aid under Public Law 480 and in this connection I may spend some time this summer testifying before Congressional Committees [for Abu Simbel]. Later I shall need letters from you, but, meanwhile, for our plans and for the sake of our presentation in Washington, we do need indication of the support of the institutions involved. Dr. Hanfmann is securing this at Harvard at the same time.[35]

Harvard President Nathan Pusey advocated for PL 480 funding for the Expedition in his June 6, 1960, correspondence to Mr. Robert H. Thayer, Special Assistant to the Secretary of State:

> Harvard has full confidence in the scholarly competence and ability of Professors G.M.A. Hanfmann and A.H. Detweiler who direct the explorations. Their work in Sardis has already yielded significant additions to knowledge and is full of promise for the years to come. Close connections between American and Turkish scholars, officials, and workmen is essential to success in this undertaking. It is reasonable to believe that such intimate collaboration over a period of years contributes much toward the growth of understanding, confidence and respect between individuals of the two nations. We would welcome financial assistance for the Sardis Exploration under Public Law 480, Section K and hope that it is possible for this to be arranged.[36]

An identical letter was sent to the Honorable Fletcher Warren, American Ambassador to Turkey. Malott, too, followed up with a letter to Ambassador Warren:

> I would like to add my endorsement to the application for financial assistance [for the] "Archaeological Exploration of Sardis" under Pu[b]lic Law 480, Section K. This expedition is sponsored by the American Schools of Oriental Research, Cornell University, and the Fogg Art Museum of Harvard University.
>
> Research projects such as this would seem to be an opportunity for cooperation between Turkish and American scholars, and additional support from governmental sources would insure an even more favorable impression. It would seem to me that the more we could emphasize our cultural interests, the greater our eventual influence would be on that part of the world.
>
> Cornell intends to continue supporting this enterprise and would welcome assistance in increasing the effectiveness of this undertaking.

On June 20, 1960, Detweiler, in his capacity as President of ASOR, wrote to Ambassador Warren:

> The American Schools of Oriental Research act as the chief sponsor for the "Archaeological Exploration of Sardis," guaranteeing the professional performance of the expedition. I investigated the possibilities of the site before obligating my organization, of which I am president. I was so impressed by the promise of the city that I brought in my own university, Cornell, and put my professional reputation on the line when I assumed the responsibilities of Associate Director.
>
> In short, I believe that Sardis presents a great opportunity for American scholarship to show Turkey what can be accomplished in the field of Archaeology. If funds were available, it would be possible to inaugurate a cooperative enterprise with Turkish scholars and, in addition, to train people from both countries in fieldwork.
>
> The United States is almost the only large power that does not express its interest in scholarly affairs by financial support. I would therefore strongly urge that financial assistance for the Sardis Exploration be made available under Public Law 480, Section K.[37]

As discussed previously, Abu Simbel and JFK's speech tend to be seen as the catalysts for PL 480 funds and archaeology. But the dates suggest that it was, in fact, Sardis, and likely Detweiler and Hanfmann, who may have first realized the potential of dedicating PL 480 funds to archaeological research, likely via their networks in Washington, D.C.

In Detweiler's personal files archived at Cornell University is a copy of Myron Smith's June 9, 1961, statement to the House of Representatives, Committee on Foreign Affairs, Subcommittee on State Department Organization and Foreign Operations. The statement was to support the proposal of the 1961 Mutual Educational and Cultural Exchange Act. Mr. Myron B. Smith worked in the Library of Congress; his interests were in Near Eastern studies and Islamic Architecture. His fieldwork in Iran in the 1930s earned him a position with US intelligence in World War II, and thus, he became indoctrinated into how the US government functioned. His carefully worded statement, especially its clearly articulated policy goals, reflects this training. He advocated for increased cultural exchange between the United States and the Near East. Upon concluding, he referenced a memorandum of a conversation he had had on April 7, 1958, with Mr. Keith E. Adamson, then Public Affairs Officer at the US Consulate General in Istanbul, entitled "American Institutes for the Near East."[38]

Among the topics discussed were archaeological missions, and among the solutions for funding (broadly) was point (d)(1): Public Law 480 funds, from sales of surplus agricultural commodities. Hanfmann wrote from Istanbul on April 12, 1958, of his preparations for the excavation house, and he lamented inflation—prices were 10 percent to 15 percent higher than expected.[39] Funding was clearly on his mind during his visit to the US Consulate.

As described earlier for Abu Simbel, Louisiana Senator A. Ellender, who served as Chair of the Senate Committee on Agriculture from 1951–1953 and 1955–1971, pushed Luke Battle to explore excess currency for financing archaeology in Egypt.[40] Discussions and planning for the 1954 Agricultural Assistance program, PL 480, took place under Ellender, giving him keen insight into not only how the legislation and funding were planned, but also how funds sitting idle might be used for other purposes, purposes with the US name attached to them.[41]

As discussed in Chapters 3 and 5 of this book, the year 1954 was crucial for US-Turkish relationships. Discussions over Turkey's financial stability were rocky, owing to the World Bank's negative rating, yet also hesitations of the US commitment to Turkey. This was dictated in part by economics, but also the US allegiance to the NATO base in Izmir. Özcan's research shows that between 1954 and 1958 the Turkish government was granted $153 million USD through PL 480.[42]

When Detweiler and Hanfmann advocated for use of PL 480 funds in 1960, there were sufficient resources available. Members of US Congress, including Ellender, were aware of how PL 480 could assist archaeological projects. I suggest that this may have been part of the reason (if not the impetus) for Kennedy's suggestion in 1961 to tap such funds. Here I recall JFK's statement for the opening of the 1961 Izmir International Fair quoted at the outset of Chapter 3 as well as the PL 480 funding for the US pavilion.

Having reviewed proposals submitted in 1960 and 1961 by the Sardis team, in 1962 the US Department of State awarded Harvard a grant for $180,000 under PL 480. These funds had to be used to advance "mutual understanding" at Sardis per the regulatory guidelines for the expenditure of excess funds in-country. In response to the deliverables from these funds, Hanfmann wrote, "In the short span of five years, the Sardis project has become a large-scale operation comparable in scope with enterprises like Ephesus and Pergamon, which have been in progress for decades."[43] Hanfmann envisioned that the practice of archaeological research would be as much about understanding the past as it would be about the present.

The first large capital project was the construction of the main building of the Sardis excavation house, shown in Figure 4.6:

> This mission was launched in a close cooperation with the Turkish Department of Antiquities of the Ministry of National Education. Thus, an excavation house is being constructed jointly by the expedition and the Ministry. After the end of the excavation, it will become Turkish Government property to be used as a museum for the antiquities excavated.[44]

Support for the second phase, which included more residential rooms, conservation areas, a depot, and a "Trainee House" for Turkish colleagues and students came from PL 480 funds.[45]

This foreign footprint served not only US interests but also Turkish goals. Turkey's five-year plan (1963–1967) called for the dedication of greater resources to agriculture and also to infrastructure for tourism enhancement. Hanfmann's proposals coincided with this wave of support.

Resurrection

With a primary excavation house complete, Hanfmann and Detweiler began to craft an image to raise further funds for excavation and restoration. Stories of the archaeological work ran in the *New York Times, London*

FIGURE 4.6 Main excavation house at Sardis, 1959. Courtesy of the Archaeological Exploration of Sardis.

Times, Boston Globe, and *Washington Post*. More extensive narratives were published in *Türk Arkeoloji Dergisi*, the *ICOMOS Bulletin*, and various outreach publications of Harvard and Cornell universities. In late 1961 it became clear that infrastructure for tourism could not wait. In a letter to the General Directorate, dated June 30, 1962, Hanfmann proposed an initiative:

> In principle, I feel there is much to be said for the solution that you propose, namely that of having a museum near the main highway. There is one that comes to my mind in this connection. As you know, the Roman gymnasium (so-called Bedesten) is preserved to a very considerable degree. It faces on the wonderful and impressive Roman marble avenue. Both, from the point of view of location and from the point of view of the setting of antiquities, it would make an excellent museum, if suitably restored. I wonder what you think of this possibility. . . .
>
> PS. One important matter remains. We have stated in our application to the Department of State that your Department has promised, as ineed [*sic*], it had to contribute funds toward a program of restoration and conservation of monuments. Any construction of a museum would have to be on the same basis as the programs now contemplated for Bodrum and Gordion and previously for Ephesus—that is, it would have to be a joint enterprise with some funds coming from your Department.[46]

Hanfmann was responding to Rüstem Duyuran's letter of June 27, 1962. Mr. Duyuran had described the issue of visitors and the inappropriate location of the current excavation and research complex as too far from the road. For this reason, Duyuran proposed a museum near the main road, rather than at the excavation house. Hanfmann endorsed the new location, yet not the plan for a "normal" museum. He proposed to reconstruct the Marble Court and the Bath Gymnasium on a monumental scale. The vision came from Detweiler and that of his student, Stephen W. Jacobs, and has roots in what had been done earlier in Athens with the Stoa of Attalos (1952–1956) and at Ephesus with the Temple of Hadrian (1957–1958), and, more specifically, what was being planned for Abu Simbel, described in 1960 by Desroches-Noblecourt as an open-air museum.

The Smithsonian supported this plan for Sardis, writing to the US Embassy in Ankara: "To summarize, I believe it is correct to say that the work at Sardis is now at a critical juncture. There yet remains new knowledge to be discovered from the excavations, but the principal task from now on will be restoration of the site, so that it is left as an Outdoor Museum."[47] Detweiler approached Cornell University President Malott for another letter to Washington to secure additional funds under PL 480.[48] At the same time, Battle had decided to pursue PL 480 funds for Abu Simbel, per the recommendation of Senator Allen Joseph Ellender.

The first stage for reconstruction at Sardis was research and full documentation of the buildings. Jacobs spent three seasons cataloguing the finds that then served as the blueprints for the restoration. Hanfmann and Detweiler convinced their colleagues that enough of the structures (Marble Court, Bath Gymnasium, and Synagogue, discovered in 1962) were present to justify as full a reconstruction as possible. This decision followed contemporary protocols closely. *Anastylosis* was to be avoided unless a significant portion of the building was in place, as suggested by the 1931 Athens Charter. The conversations through the 1950s and early 1960s on this topic framed the protocols embedded in the 1964 Venice Charter.

Hanfmann pushed the limits. He framed the project as a vision, not as a fact: "The structure as restored suggests what the Romans might have built, rather than what they must have built."[49] Harvard-trained architect Mehmet Bolgil endorsed this approach: "[t]he general aim of the restoration was to recreate the grandeur of Imperial Roman architecture and give the visitor an opportunity to discover first hand the intricacies and secrets of the masterful architectural detailing of that period without having to decipher the technical drawings in textbooks."[50] Fikret Yegül, too, outlined the difficulties in restoration, presenting both possibilities and limitations: "It should be understood that the objective of historical reconstruction is not the recreation of the institution, but only the physical reality of the building."[51] His detailed perspective drawing of the Marble Court, illustrated in Figure 4.7, gave people an idea of what a building in ruins, as shown in Figure 4.8, could look like once resurrected.

Correspondences between Hanfmann, Detweiler, and the architectural team (Jacobs, Yarnell, and Yegül), and others reveal that this vision (and the final product) did not come easily. Dozens of studies were done.[52] Design meetings took place after dinner with the faculty, students, and consultants involved (Yegül, personal communication, 2017). The team took great efforts to consider the immediate challenges (and opportunities) at Sardis. In addition, international standards in conservation, archaeological science, and landscape studies shaped how the work progressed, as described by Yegül:

> The choice of the building or the type of building depends on the site, the conditions of the structure itself, its historical and professional significance,

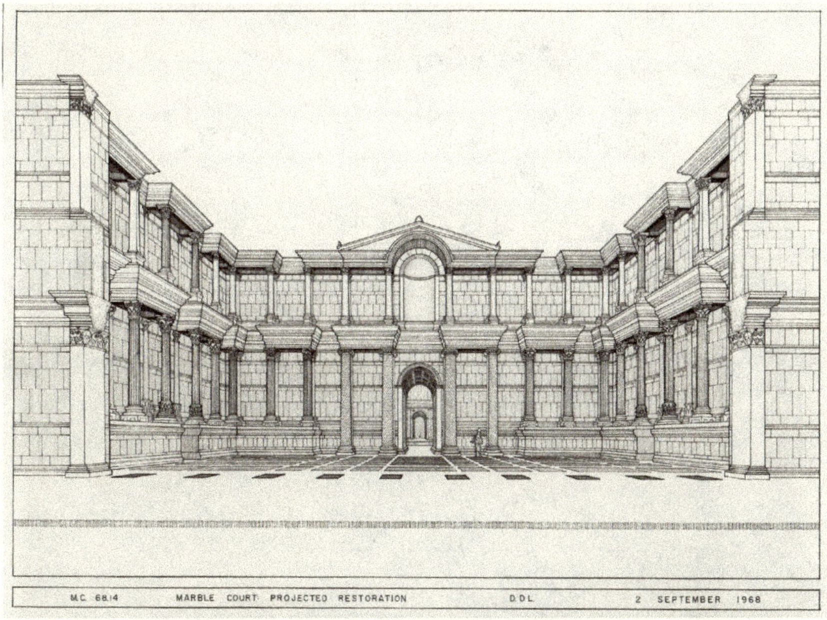

FIGURE 4.7 Reconstruction perspective of the Marble Court, 1968. Courtesy of the Archaeological Exploration of Sardis.

> technical feasibility, and present-day usefulness. The extent to which a chosen building is to be reconstructed and the manner in which it is done are indeed central issues to the problem.[53]

The Sardis team agreed on the following protocols:

> 1) Original materials were used as much as possible. The dismantling, reinforcing and rebuilding of original parts were done according to the original scheme which had been recorded and preserved by drawings and photographs. 2) Where new materials were used, the change was expressed in a way that approximated old materials in order to avoid strong contrasts. Color and texture were utilized to give the new elements a suggestive character.[54]

Yegül cautioned against attempting to reconstruct the "original" precisely because research precluded 100 percent accuracy when determining when and where each individual element had been incorporated into the building:

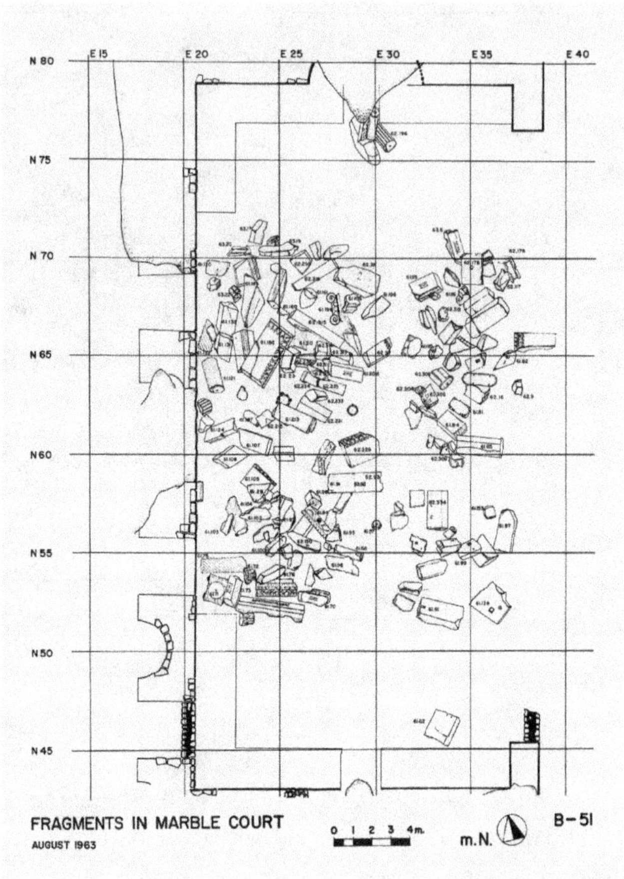

FIGURE 4.8 Work-sketch showing findspots of the fallen architectural fragments in the Marble Court, 1963. Courtesy of the Archaeological Exploration of Sardis.

> A certain consideration of principles and policies comes first: how faithfully will the reproduction imitate the original? If the objective is reproducing a facsimile, the enterprise might border on plain forgery. The problem is in finding the right method which will allow a definite distinction between what is new and what is old without creating crude clashes. Acknowledging the difference between the two constructions becomes a question of honesty. This further protects the scholar against risky professional commitments since his intentions are exposed as tentative and conditional.[55]

This project was a monumental proposal for Turkey, as Hanfmann wrote: "Together with the Stoa of Attalos in the Athenian Agora and the facade of the Library of Celsus in Ephesus, this may be the most

FIGURE 4.9 The first architrave of the screen colonnade is moved into position, 1970. Courtesy of the Archaeological Exploration of Sardis.

extensive reconstruction work undertaken in the Aegean in the last few decades."[56] He believed it was the largest project north of Baalbek.[57] Detweiler's training and experience in the reconstruction projects in Jaresh, Jordan, too, provided needed expertise in realizing a project of this scale from initial survey and documentation to resurrection of an ancient façade.[58] The next process was organization and machinery necessary to put the pieces together, as shown in Figures 4.9 and 4.10. A closer look, however, reveals a project fraught with tensions and entanglements primarily resulting from the lack of foresight in US diplomacy in the Mediterranean.

FIGURE 4.10 Reconstruction of the façade of Marble Court, 1970. Courtesy of the Archaeological Exploration of Sardis.

Reconstruction

Hanfmann recognized at the outset that good policy meant working closely with the respective cultural agents. The element of collaboration in the Sardis reconstruction program provides a window into his commitment to professional integrity and best practices. Hanfmann took great pride in the training program supported by PL 480: "In this country, the grant of funds in Turkish currency by the Department of State (1962–1965) was a pioneering venture; its greatest benefit was a regular training program for Turkish student trainees, several of whom became valued colleagues in the Sardis work."[59] These people included M. C. Bolgil, F. K. Yegül, M. T. Ergene, and T. Yalçınkaya. In total, at the close of Hanfmann's tenure at Sardis in 1975, he was proud that close to seventy Turkish professional partners and students—in addition to hired staff and excavators—had been trained.[60]

The social and political dimensions of the Expedition are presented in *Letters from Sardis*, a collection of the newsletters sent to the sponsors of the Expedition.[61] These published letters, however, neglect the turbulence of the times and the lamentable actions of the US Department of State regarding decisions to pull future support from PL 480, offering nothing in return. Recall that on September 4, 1963, the Senate Appropriations Committee struck down all funding for archaeological projects as official policy moving forward. While Abu Simbel was well underway, Sardis had not yet solidified a full contract, only a verbal commitment. The initial PL 480 grant for 1962 thru 1965 covered the archaeological research and additional documentation; funding for that was secure. But funding for the Expedition's commitment to the final phase, the actual restoration of the Bath-Gymnasium and Marble Court complex, had not yet been secured, even though the US Embassy and the Smithsonian had indicated that support would be forthcoming through PL 480, Section K. To learn that it was *not* forthcoming was devastating.

First, Hanfmann rallied support from primary sponsors—Cornell and Harvard—and colleagues in other institutions working in Turkey, such as those at the University of Pennsylvania, University of Michigan, and Columbia. Second, he organized the AIA and ASOR to lobby for US foreign policy that included support for archaeology. Third, he focused on Washington and the Smithsonian Institution's Office of International Activities, which distributed PL 480 excess funds. Finally, he wrote to members of US Congress and staff at the US Department of State. The common theme throughout all his correspondence was the importance of promoting cultural relations between the United States and Turkey.

The efforts began in earnest in the winter of 1964, just five months after the shift in policy and less than two months after the assassination of President Kennedy. AIA President Johnson requested a transcript of "the hearing of Luke Battle and his aides before the Subcommittee on Appropriations (the McClellan Committee)." The report was devoid of specifics regarding archaeology. As Johnson wrote to Hanfmann in a letter of February 21, 1964, "I suspect that remarks hostile to or unfavorable to certain programs have been edited out or toned down. This is what I hope to clarify. . . ."[62] In Washington, Johnson met with Mr. Hummel, then the

Acting Assistant Secretary for the Bureau of Educational and Cultural Affairs.

Johnson again wrote to Hanfmann on March 2, 1964: ". . . the outcome with Mr. Hummel was negative, even with the reinforcement of your remarks. It is not now the policy of the State Department to support archaeology."[63] Mr. Battle and Mr. Hummel held firm. Johnson's closing to Hanfmann reflects the dire situation in Washington:

> Ergo, we—you, I, and our grandchildren—have before us a colossal lobbying job, to tell the senators and congressmen what archaeologists do and why and how they do it. I am hammering out a statement and I shall steal ideas from anyone I can, including yours. . . . In the meantime, if you take any drastic action, or have any convulsive ideas, keep me informed.[64]

The US Department of State would meet commitments at reasonable budgets for the foreseeable future. These included the American Research Center in Egypt and the American Research Institute in Turkey, but not future archaeological fieldwork.

In a February 23 1964, correspondence, Hanfmann sent Johnson a long, detailed letter with talking points to convince the US State Department of the value of archaeological endeavors. He laid out the key elements for how archaeology contributed to public life in contemporary society. He stressed tourism and economic development, and the expectation of good faith and solid, reliable cultural diplomacy as one component of a nation's foreign policy. He explained that such projects should not be one-offs, but rather had to be long-term commitments. The templates of those followed by the Austrians, French, Germans, Dutch, and Italians were highlighted, as was President Kennedy's explicit 7 April 1961 endorsement of archaeology and preservation as global goods and moral priorities.[65]

The negative tenor from Washington also led Hanfmann to write to Ernst Kitzinger, Director of Byzantine Studies, Dumbarton Oaks, further grappling with the reality of the new US policy. Johnson had dug into as much detail as he could find from the legislative and executive branches. He even "inspected the published transcript of a hearing of the Subcommittee of the Senate Appropriations Committee." Hanfmann concluded that if "certain remarks allegedly made by certain Senators were actually made, they do not appear in the published record." In fact, Johnson could find *no* record of archaeological programs or archaeology having been discussed. Hanfmann hoped to explain to US authorities "the merits of archaeology as an instrument of the cultural policy of the United States."[66]

For Hanfmann the conservation and restoration of the past created opportunity for the future, and it also validated the expenditures to date, as communicated in his letter to Battle on April 22, 1964:

> The Archaeological Exploration of Sardis needs to receive continued support in order to attain the scientific and cultural objectives described in the original application. The next phase of the program will show increased emphasis on matters dealing with the permanent benefit of the program to the historical and artistic heritage of Turkey; the restoration of monuments and conservation of objects are of considerable concern to the Turkish authorities.

He lamented that Mr. John Forbes, Chief of Turkish, Greek, Iranian, and Cyprus programs at State, placed "all archaeological projects in the category of low priority." Hanfmann closed the letter with the following concession:

> In deference to your expressly stated current policy, Harvard College and the participating institutions will refrain from filing an application for renewal of the grant SCC 29 543 at this time. Should a change in policy occur and funds become available, we trust that you will immediately notify the College.[67]

Those in Washington who may have privately supported the Sardis cause could not do so because their efforts focused on saving the public US promise to Egypt. On May 28, 1964, Warner, consultant to the Secretary for International Activities at the Smithsonian, wrote to Hanfmann:

> [T]he Nubian monuments item may come up any day now in the State Department's Senate hearings. Dr. Ripley has been meeting with and writing to key Senators on the Appropriations Subcommittee, and we have our fingers crossed. It won't be easy because the house absolutely disallowed our $12 million foreign currency pledge and so now Senators have to be found who are willing to sponsor an amendment to the current House appropriations bill.
>
> On our prospects for administrating PL 480 funds in general, I can only report that since our phone conversation we have gotten absolute encouragement from the State Department seeking authority on our own by Congressional amendment. Perhaps encouragement isn't the right word;

let's say we have gotten nihil obstat from State, for archaeological projects only.[68]

Negotiations for PL 480 funds to Abu Simbel were handled by Battle, who displayed none of the same energy he had for Egypt for Sardis. Perhaps his insider track (once through Kennedy) was no longer available. Perhaps his personal interests in Egypt were far greater than those in Turkey, as reflected in his subsequent posting as the US Ambassador to Egypt.[69]

In his final report to the Bureau of Educational and Cultural Affairs in 1966, Hanfmann emphasized his disappointment with the trajectory of US Cultural Diplomacy. To capture his rhetoric and passion, I have chosen a particularly long section from one of his letters:

> The Directors, at least, of the Sardis program had considered this application for a grant from the United States Government as a precedent-setting step toward the recognition by the United States Government that archaeological inter-cultural programs of the kind represented by the Sardis program are legitimate and important means of cultural and educational policy of the United States. It was their hope that the program could be carried forward on the renewal basis to a definite and impressive conclusion.
>
> From the point of view of the intent of Public Laws referred to, the Sardis program and other similar archaeological programs have successfully fulfilled the double objectives: to promote the cultural relations with Turkey within a stipulated period of time, and to put to productive use foreign excess currency funds vulnerable to inflation.
>
> From the point of view of effective long-range cultural policy, however, this type of support, which depends on availability of foreign excess currency funds, is not the best conceivable means to create an image of the United States of America as a cultural leader. We would emphasize that for more than a hundred years several European (and more recently, Asiatic) countries have maintained continuous archaeological programs in Turkey and the Middle East generally. As a result, archaeological programs are one of the yardsticks by which educated society in these countries measure the cultural status and seriousness of cultural efforts of foreign countries.
>
> In the case of the Sardis program, a change of policy in the Department of State (September, 1963), and the depletion of foreign excess currency holdings in Turkey precluded the renewal of the grant to a program, which has been yielding satisfying and beneficial results.
>
> It is much to be hoped that the United States Government, through a suitable agency, will find means to carry on such programs on a more sustained

> basis. . . . Our restoration projects, initiated during the grant period, are demonstrating modern techniques and standards, and are a matter of great interest to the Turkish government. . . . [The Sardis program] has spread among wider Turkish circles the understanding for the necessity to preserve Turkey's historical heritage, and has demonstrated modern methods and means for doing so.
>
> The head of the French mission stated, upon seeing our activities in 1965, "One works here in a manner truly worthy of America.". . . Ambassador R. Hare is on record with the opinion that the Sardis program and similar archaeological programs were of high value because they benefited the host nation, while respecting its dignity, and that lira for lira such programs represent some of the best returns on foreign currency grants.[70]

Hanfmann did secure funding from the Turkish Ministry of Education and the US NEH to complete the restoration project at a respectable level given his promises to the Republic. The funds were not enough to continue the joint training of Turkish and American students, an aspect of the Sardis program that Hanfmann had taken great pride in establishing, including integration with local schools in nearby Salihli. He referred to this cut as "the first casualty of the changed situation." The US State Department showed no interest in cultivating the program and infrastructure it had funded, which Hanfmann described as "one of the best equipped research field stations in the Middle East."[71] Given his own disbelief, he had an even greater task in explaining the situation to his Turkish colleagues.

Hanfmann wrote to William Warner at the Smithsonian on February 26, 1965, "A sudden retraction and abandonment of the program which will occur if no public funds are made available may have unfavorable and unfortunate repercussions in Turkey."[72] It wasn't merely research and infrastructure, but credibility with the Turkish Department of Antiquities and Ministry of National Education. Hanfmann was firm: "The ultimate success or failure of the program depends to a considerable extent upon our ability to leave behind us not a chaos of unintelligible ruins but an image of what an ancient city was like—a site attractive in appearance and intelligible to the public. . . . When completed, this complex will be one of the most imposing ancient buildings to be seen anywhere in Turkey."[73] The US State Department refused to consider new options for funding.

Another staggering blow came in 1967 when Cornell University President Perkins withdrew the annual contribution of $10,000. Cornell had reconsidered the relationships between academic research and development, especially given the rising tensions with Cornell programs operating in other parts of the world (e.g., Jordan and Peru).[74] For Perkins, research at Sardis was moving toward a development project beholden to the whims of the US and Turkish governments.

Even with detrimental policies from the United States, the Sardis team did meet their baseline commitment to restoration. At the time, it was believed to be the largest program ever undertaken in Turkey. In pursuing funding and realizing the project, Hanfmann and Detweiler had fulfilled the goals of the Mutual Exchange Act under PL 87-256. The original US commitment to Sardis offered returns in spades. It became a model for future projects including the reconstruction of the Library of Celsus (1970–1978) at Ephesus by the Austrians, American activities at Aphrodisias and German-funded reconstruction at Pergamon. All three places remain major points on tourism routes for western Turkey.

Discussion

On November 16, 1963, US Congress voted to support funding for Abu Simbel. The public presentation of US support for Abu Simbel was not the first, as many believed, but rather the last breath of serious US commitments to heritage abroad. Abu Simbel was not a rallying call for a global policy, but rather a political conquest for Luke Battle. His approach, while not necessarily duplicitous or malicious, favored political momentum and congressional alliances over a commitment to global heritage policy. The inexcusable irony, however, is that Battle's solution (i.e., PL 480) had in part been forged by a Public Affairs Officer at the US Consulate in Istanbul (Keith E. Adamson) and a member of the US Library of Congress (Myron Smith), and yet, Battle refused to pursue the same vigor for archaeology in Turkey at Sardis, or frankly anywhere in the world, even with his amplified prestige as US Ambassador to Egypt.

The history of funding for Sardis exposes the connectivity of seemingly disparate decisions made in closed-door sessions in Washington, D.C. With Soviet tensions high in Egypt as well as Turkey—and obsolete nuclear-tipped missiles placed in Izmir as a symbolic show of force against the Russians—there was ample reason to support the cultural dimension in Turkey, just as the United States was doing in the early 1960s in Egypt. This case study shows that despite being the poster-child project for the 1964 Venice Charter, the formation of ICOMOS, and the 1972 World Heritage Convention, Abu Simbel was in fact a dire blow to US-supported archaeological research abroad. The United States turned away from the field archaeology as a component of its diplomatic toolkit. At Sardis we can now document how these decisions reverberated not only in the center of Sardis proper but also at a regional level. Without support, the largest monumental cemetery in Turkey, perhaps the world, Bin Tepe, was abandoned and with it an entire research program focused on the Gygaean Lake Basin. In its place, US aspirations of agricultural development were unleashed, establishing a different agenda for landscape management, the topic of Chapter 5.

CHAPTER 5 | Organic Lives

The situation at Bin Tepe represents a real problem. This unique complex is disappearing fast under the onslaught of mechanized agricultural machinery.[1]

—GEORGE HANFMANN, Director of Sardis, 1965

Our brand is named after the richest king of antiquity, Croesus, father Alyattes, who was also one of the longest reigning kings in the history of mankind.[2]

—ALYATTES, an olive company located in the heart of Bin Tepe, 2015

ONE AFTERNOON IN 2011, a village elder in his seventies recalled the trajectory of his life over tea. His family had initially practiced transhumance, yet once they had "settled" in the early 1950s they had relied on a mixed agro-pastoral economy, which had given them sufficient resources to send him to school. He rode a camel, fording rivers along the way. He also recalled when they got their first tractor in the early 1960s, "from the Americans." He then smiled and said, "Marshall gave me my land." Today this family is among the most prosperous not only in this small village, but also in the region of Bin Tepe in western Turkey.

In 1989 the village partnered with the German firm Rapunzel Naturkost. Through contract farming arrangements, Rapunzel purchased organic fruits. In the early 2000s, families expanded into organic olives, and in so doing, took on more risk. Contract farming with Rapunzel was no longer an option given EU regulations for competing crops. Turkish olive oil contends with Spanish, Italian, and Greek olive oils. Fifteen years later, this risk has proven to have been a prudent one. Many families have come into unprecedented wealth, reflected materially in homes, cars, and farm equipment, but also in education. The man who rode a camel to school at age ten has seen his two granddaughters, who began first grade in a one-room village schoolhouse, graduate from universities in Izmir and go on to study in the United Kingdom. From a development perspective, this family's livelihood resides in the redoubling success of the family farm. It is a Marshall Plan success story. But most of their orchards and vineyards are illegal because they are located in the 72-square-kilometer area of Bin Tepe that is designated the highest level of protection for archaeological zoning in Turkey. Technically, agricultural activities are forbidden.

This chapter investigates intersecting paradigms and policies of archaeological practice and agricultural development in the region of Bin Tepe, placed on the Turkish tentative list for UNESCO World Heritage sites in 2013: *The Ancient City of Sardis and the Lydian Tumuli of Bin Tepe*. I explore also the abutting wetlands of the Marmara Lake Basin, registered as a Protected Wetland Area by the Turkish Ministry of Environment and Forestry in 2008, as an Important Bird and Biodiversity Area (IBA) by the International Union for the Conservation of Nature (ICUN) in 2009, and a national ecological protected area in 2017.

"A thousand mounds" in Turkish, Bin Tepe is known for its rolling landscape of tumuli, man-made earthen mounds that cover a burial chamber, usually hewn from natural bedrock or built of quarried stone.[3] At one time at least 140 tumuli dotted this landscape, making it the biggest tumulus cemetery in Turkey. The tumulus of Alyattes, fourth ruler of the famed Mermnad Lydian dynasty, is the largest tumulus in Turkey: at the base it is 355 meters in diameter and 63 meters high. The closest landscape that rivals Bin Tepe in size and character is that of Giza with its Pyramid of Khufu (230 meters at the base and 146 meters high). The archaeological zone of Giza covers 160 square kilometers. In 1979 it was placed on the World Heritage list, *Memphis and Its Necropolis—The Pyramid Fields from Giza to Dahshur*. While the Pharaonic past has been deemed of Outstanding Universal Value (OUV), the landscapes of Bin Tepe that captivated Herodotus have yet to share the same status, and they are subject to erasure from agricultural activities.

The hyperconnectivity of expanding organic agribusiness threatens Bin Tepe. At the core of this analysis are the incongruities within preservation approaches, development agendas, and expansionist corporatism. This story explores venerable ties to legacy Ottoman families, foreign diplomatic and development policies, the rise of neoliberalism, and the mixed blessing of transnational corporations (TNCs) in the spheres of organic agriculture in

Turkey. Building on the US and Turkish water diplomacy programs of the Aegean TVA discussed in Chapter 3, and the failed promises of the US cultural dimension at Sardis discussed in Chapter 4, this analysis explores the entanglements of heritage and agricultural policies on a landscape that has been subjected to intensive industrialization driven by capitalist programs with the primary goal of cultivating entrepreneurial farmers. This has been at the expense of archaeological and natural heritage.

Even as authorities balk at the harmonization of agricultural, water, and archaeological policies, farmers masterfully manipulate heritage zoning to avenge personal vendettas and leverage contract farming relationships with European partners to secure loans, which banks would normally not offer for lands in archaeological areas.

To make the situation an even richer case study, the permits for archaeological research in this area are held by two American teams: the Kaymakçı Archaeological Project (KAP) 2013 to present (also responsible for the Central Lydia Archaeological Survey (CLAS) 2005–2014) and the Archaeological Exploration of Sardis (1958–present). Neither project, however, currently holds jurisdiction for excavation and management of Bin Tepe or the surrounding Marmara Lake Basin. Management of these resources falls to the local Manisa Museum of Archaeology and Ethnography and the regional Second Protection Council in Izmir, neither of which has been able to enforce regulations pertaining to preservation or wise-use management. As I show in this chapter, the result has been that archaeologists from all sectors are paralyzed to effect change against agricultural, hunting, and fishing lobbies precisely because these activities ensure self-legitimatization for local communities and ultimately highly valued regional economic stability. This study exposes the limits of change offered by national preservation laws and international prestige tags such as UNESCO's World Heritage designations and the IUCN's IBA zones.

Zoning Wars: The Legislative Gauntlet

UNESCO's initial guiding statement, "Peace in the Minds of Men,"[4] provided a platform for not only key cultural and natural heritage policies of the post–World War II order, but also agricultural paradigms. Scholars of heritage studies often applaud the establishment of the 1946 International Council of Museums (ICOM), the 1948 International Union for the Conservation of Nature (IUCN), the 1954 Hague Convention for the Protection of Cultural Property in the Event of Armed Conflict, the 1964 Venice Charter and the International Council on Sites and Monuments

(ICOMOS), the 1970 Convention on the Means of Prohibiting and Preventing the Illicit Import, Export and Transfer of Ownership of Cultural Property, the 1971 Ramsar Convention on Wetlands, and the 1972 Convention concerning the Protection of the World Cultural and Natural Heritage. Yet what of parallel initiatives, both international and national, with respect to agricultural development?

In fact, over the years, agricultural and water management programs have been far more handsomely supported than those related to heritage. A social history of policies for Bin Tepe reveals how the implementation of the first Turkish five-year development plan and Seed Law in 1963 shifted the quality of life for rural communities through programs and loans from the World Bank and the United States Agency for International Development (USAID). While the trajectory of TNCs and the neo-liberal Turkish agricultural policies of the late twentieth and early twenty-first centuries have been less invasive in the Marmara Lake Basin than elsewhere in Turkey, partnerships with Rapunzel and the growing demand for "high-value" organic foods (olive oil, nuts, fruits) still damages cultural and natural heritage.

The Layers of Bin Tepe

The places in question are the earthen mounds or *tumuli* of Lydian and Persian-era Bin Tepe and the archaeological sites nestled between them, including among the earliest (Lower Paleolithic) sites of human occupation in Anatolia.[5] The tumuli are testament to the prominence of this landscape in the sixth and fifth centuries BCE and have been the subject of extensive scholarship.[6] Their monumentality ensured they were admired and explored by Hellenistic through Ottoman communities in the region. And, they were the muse of early travelers and archaeologists.[7] In May of 1880 George Dennis was issued a *firman* (i.e., formal permission) for excavation in Bin Tepe. Shortly thereafter he moved to purchase the Temple of Cybele.[8] While tumuli and ancient monuments may have inspired and continue to evoke special interest in Bin Tepe, the region has a much more extensive recent history that is relevant to understanding the preservation quagmire.

Ottoman-Era Communities

The earliest documentation of a village in Bin Tepe on a map comes from Giovanni Battista Borra's sixteenth-century map illustrating the location of key monuments, a cultivated plain, and at least one village. Figure 5.1

FIGURE 5.1 1750 Illustration of Gediz plain from Sardis. Giovanni Battista Borra. Courtesy of the Yale University Archives.

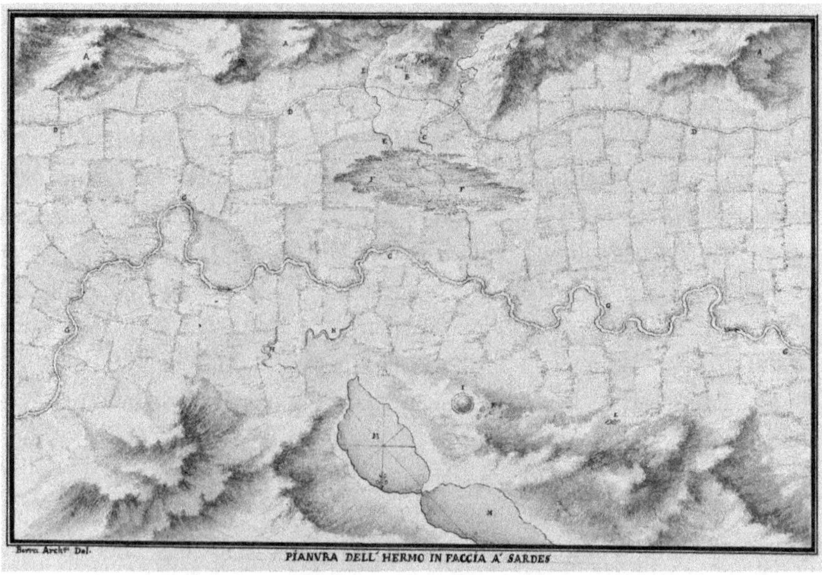

FIGURE 5.2 1750 Illustration of Gediz plain. Giovanni Battista Borra. Bird's Eye view of Bin Tepe and Marmara Lake. Courtesy of the Yale University Archives.

FIGURE 5.3 1836 The Gygean Lake and the Place of the Thousand Tombs, Asia Minor. Illustration from *Constantinople and the Scenery of the Seven Churches of Asia Minor illustrated., With an historical Account of Constantinople, and Descriptions of the Plates, London/Paris, Fisher, Son & Co.* (1836-38), by Robert Walsh and Thomas Allom. Courtesy of the author.

is Borra's illustration of the plain from Sardis with Bin Tepe and Lake Marmara in the distance. Figure 5.2 is a bird's-eye representation of field boundaries. Borra's cultivated plain corroborates evidence found in the remarks of the fourteenth-century writer Ibn Battuta regarding the region's viniculture.[9] Clive Foss and Feridun Emecen, too, discuss Greek communities working these vineyards under the auspices of wealthy Ottoman estates (*çiftliks*).[10] Thomas Allom's illustrations of the region, especially the Gygean Lake and Sardis, are well-known images from the early nineteenth century (Figures 5.3 and 5.4). Photographs taken from the Acropolis at Sardis over the Gediz valley by Olmstead in the early twentieth century also confirm the continued practice of cultivation and are shown in Figures 5.5 and Figure 5.6.

In Bin Tepe proper and along the southern shore of Lake Marmara there were a number of nineteenth- and early-twentieth-century communities. Russians held the fishing concession for the lake and had a small settlement;[11] a Greek community farmed nearby lands; a hunting lodge was kept by the Levantine Giraud family of Smyrna; a more substantial lodge of the

FIGURE 5.4 1838 The Acropolis of Sardis. Illustration from *Constantinople and the Scenery of the Seven Churches of Asia Minor illustrated. With an historical Account of Constantinople, and Descriptions of the Plates. London/Paris, Fisher, Son & Co.* (1836-38), by Robert Walsh and Thomas Allom. Courtesy of the author.

FIGURE 5.5 Sardis, View of Gediz Plain from Top (of the Acropolis), 1902. Courtesy of the Oriental Institute, University of Chicago.

FIGURE 5.6 View of Sardis. Plain looking North toward tombs of Lydian kings, 1908. Courtesy of the Oriental Institute, University of Chicago.

Karaosmanoğlu family was situated nearby on a high bluff; and transhumant Yörük, including the Tekeli and Karayahşi tribes, had seasonal settlement locations in Bin Tepe, part of their extensive grazing lands between the *yaylas* of Boz Dağ and those of Gur Dağ.[12] According to the research of Yuzo Nagata and recent oral histories, Bin Tepe could support 1,500 sheep and/or goats for approximately two months.[13]

In 1841 Yetim Ahmet Ağa Karaosmanoğlu established the Bağırgan estate. Just east of Bin Tepe, the location for the Bağırgan was chosen for its proximity to prime agricultural land and access to the hub of Adala as well as the emerging center of Salihli. Incentives under the 1858 Ottoman Land Code propelled changes that allowed the estate to prosper, helped in part by the expansion of railroads to Smyrna's growing economic base. The Bağırgan explored new approaches and technologies for intensive farming, management of livestock (cattle, especially), and land reclamation. Primary crops were wheat, barely, chickpea, bitter vetch, and corn. In addition to these standard crops, the Bağırgan produced cotton, root dye, and tobacco for luxury markets beyond Turkey; the shipment point was Smyrna.[14]

Thus, on the eve of the Turkish War of Independence, the middle Gediz Valley represented a dynamic mosaic of cultural identities, political interests, and economic experiments. During the Greek occupation and subsequent conflict (1919–1922) the eastern extent of Bin Tepe formed the boundary between the Greek West and the Turkish East, known as the Milne Line (Treaty of Sèvres, Article 66; see Chapter 2, this volume). The region of Bin Tepe saw intense fighting in summer of 1920, when the Greek military crossed the boundary into Turkish territory. Still today the area is known as the "Bloody Pass" (*Kanboğazı*).[15]

After 1923 the Republic issued mandates for villages to be established. Practices of transhumance were to be abandoned and pastoralism discouraged. Legal parameters that responded to the 1926 Civil Code for land ownership paved the way for the 1945 Law "Providing Land for Farmers." These were among the many reasons why both the Tekeli and Karayahşi Yörük tribes settled in their seasonal camps of Bin Tepe and in theory were able to claim ownership of lands where their herds had grazed and they had kept seasonal gardens and homes.[16] The Karaosmanoğlu struggled to hold onto their power and estates. Many family members took posts in the government that allowed them to oversee development programs favorable to their holdings such as the Bağırgan (see Chapter 3, this volume).[17]

Cultural Heritage

In 1937 the Manisa Museum of Ethnography and Archaeology opened its doors, and with it archaeologists began to oversee heritage management in the province of Manisa. In 1936 and 1940, at the request of the US Embassy on behalf of the US National Park Service of the US Department of Interior, Turkey issued a list of archaeological sites that could be developed for tourism. This list was revisited in 1950. It set the stage for the 1951 legal framework within the High Council of Real Estate Antiquities and Monuments. Sardis was among the sites listed.[18] Bin Tepe wasn't.

From the start of his work in the area, George Hanfmann, the first Director of the Harvard-Cornell Sardis Expedition (see Chapter 4, this volume), recognized the value of Bin Tepe. In May 1957 and February 1958 the Sardis Expedition requested aerial imagery for survey. The proposed topographic survey and mapping initiatives were too expensive and expansive for the Turkish Directorate General of Cartography of the Ministry of National Defense to undertake; tacit permission was given to Sardis to organize it, provided that the Expedition pay for it. The Turkish Directorate General of Cartography declined the Expedition's request for aerial imagery on the grounds of national security.[19] Eventually on March 10, 1960, Süleyman Demirel, then Director General of the State Hydraulic Works (DSI) (see Chapter 3, this volume), granted the Expedition access to nine aerial images, sending them to Hanfmann. These photographs aided in designing a research program for Bin Tepe. This work included plans for excavations at (and in) Karnıyarık Tepe, one of the three largest tumuli, and a few targeted excavations (Ahlatlı Tepecik and Eski Balıkhane) along the shores of Lake Marmara.[20] In addition, Hanfmann envisioned a diachronic survey project for the Marmara Lake Basin. In his1965 funding proposal to the Smithsonian Institution he referred to this plan as an "Environmental and Prehistoric settlement survey of the Bin Tepe-Gygaean Lake area."[21]

Hanfmann's 1965 research goals, however, were not necessarily reflected in his public essays about Bin Tepe and Sardis. In 1958 he wrote for the *Daily Boston Globe*. The title "Harvard Man Finds Treasures of the World's Richest King: Ancient Secrets Bared, Come War and Death" invoked the age-old theme of archaeology and treasure. The "Battle against Bin Tepe" ran from 1962 to 1966, which included survey and extensive tunnel excavations (1964–1966) at Karnıyarık Tepe. Archaeologists were in pursuit of a pristine tomb. Experts from the coal mines in Soma were hired to construct an entry passage, shown in Figure 5.7. The interior excavation shaft had properly buttressed walls and ceilings, rail lines to move back-dirt, and systems for lighting and air supply.

A cross-section of the operation, as presented in Figure 5.8, was drawn by Crawford H. "Greenie" Greenewalt, Jr., when he was a student (he would go on to be the Director of Sardis, 1976–2008). He featured the primary access road into the area and the residential camp, a hiking path to the top of the mound, wooden beams for shoring up the tunnels, the rail track into the center of the mound, and from this point excavation tunnels, one along the crepis (retaining) wall at the base and others placed in hopes of intersecting a chamber. At the time, archaeologists believed that the chamber was that of Gyges, the founder of the Mermnad dynasty of Lydia.[22] Greenie included a caricature of Gyges in the upper left-hand corner of the drawing, referring to him by his Assyrian name, Gugu.

Given how remote Bin Tepe was at this time, the excavation and mining crews lived on site and worked in shifts around the clock. A tomb was never found. The Expedition referred to the mound as "defiant," keeping its unresolved secret. The Expedition reluctantly admitted, "T'was a good fight." While they considered to "dig the mound from above, find all chambers, and put the mound back again," the cost was too great. For this reason, they "let king Gyges sleep."[23]

FIGURE 5.7 Entry to the tunnel of Karnıyarık. Courtesy of the author.

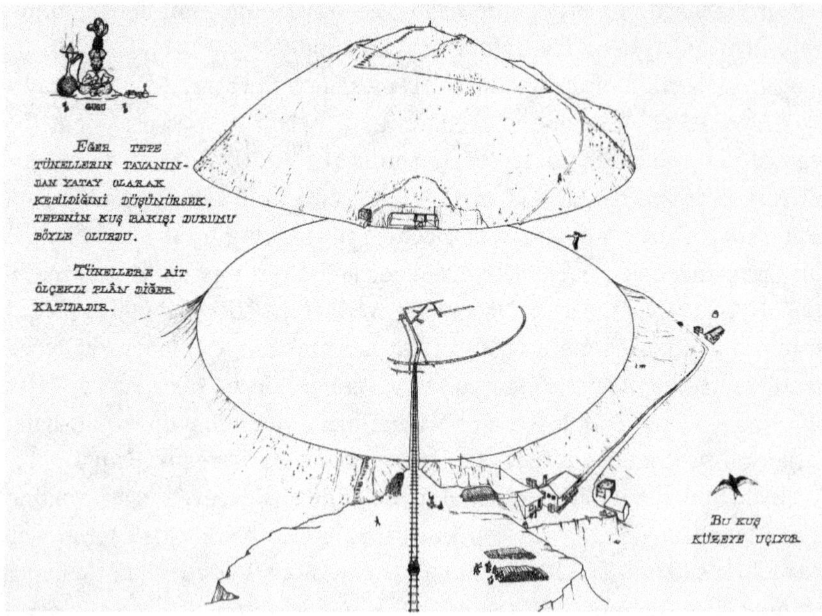

FIGURE 5.8 Illustration of excavations at Karınıyarık Tepe, Bin Tepe, 1962. By C. H. Greenewalt. Courtesy of the Archaeological Exploration of Sardis.

The premature 1964 report describing the discovery of "the great tomb of Gyges" printed by the *New York Times*, "Archaeologists Find Huge Tomb of Lydian King,"[24] fueled international attention and local fascination. Articles in the *Boston Globe* and the *New York Times* generated more awareness of the potential for treasures associated with kings than of historical relevance and preservation. During the unprecedented rains of the winter of 1966, Bin Tepe became an inaccessible island rising from the plain. This catastrophic flooding isolated the area from authorities. Elders recall a rampant and especially egregious period of plunder. US museums were among the most active buyers of items from Lydia that appeared on the private market. This included the high-profile Karun treasure case involving looted objects from the tombs of tumuli in Uşak and purchases by the Metropolitan Museum of Art (MET) in the late 1960s.[25]

After the winter of 1966, Hanfmann became increasingly concerned about the plunder of the tombs and the condition of the tumuli. In the "Restoration, Preservation, and Conservation Program" section of his 1965 proposal to the Smithsonian, he underscored the necessity of regional management. And in his final report submitted to the Bureau of Educational and Cultural Affairs, US Department of State, regarding the 1962–1966 program, he again stressed the gravity of the situation:

> The cemetery which lies some 5 miles north of Sardis is reached by ca. 12 miles of road, some of it very poor. With its ca. 100 mounds and area of 2 by 3 square miles, it is the largest mound cemetery in Anatolia. After the irrigation scheme for the Hermus valley (Demir Kopru) was put into operation, the area suffered increasingly from both agricultural activities and depredation. The directors of the Sardis program were aware of the urgency of this task but until the grant became available it could not be attempted. . . .
>
> The situation at Bin Tepe represents a real problem. This unique complex is disappearing fast under the onslaught of mechanized agricultural machinery. The Turkish State has retained property rights to only the biggest mounds. Communications are still so poor that great efforts are required to supply and provision a camp. Guards who have to be maintained at the mound of Gyges are virtually marooned in the winter. It is not clear whether the program can continue this work after the 1966 season. The minimum scientific duty is to record with or without supplementary excavation the cemetery before it disappears.[26]

The US State Department held firm in their decision not to support field archaeology in Turkey (see Chapter 4, this volume). Plans for regional management were abandoned. This decision was a humiliating blow to US archaeologists and more broadly to US cultural diplomacy in Turkey.

While the US government was unwilling to support preservation efforts for Sardis and Bin Tepe, they did provide assistance for agriculture through USAID. The expansion of the "farm to village" road systems, first implemented in the late 1940s, finally reached Bin Tepe in the early 1960s, owing to both the completion of the water works (see Chapter 3, this volume) and regional authority of the Karaosmanoğlu family. This family leveraged their waning power to influence the location of the first paved road connecting Salihli to Gölmamara (and north to Akhisar and beyond to Bursa and Istanbul). Initially the highway was slated to run north of Lake Marmara, but the Karaosmanoğlu gifted land, and thereby ensured its placement next to the Bağırgan and ultimately through the heart of Bin Tepe. The work resulted in one tumulus (No. 63.5) being sliced in half, as it remains today.

Between 1973 and 1976 the Sardis Expedition negotiated with the Director General of Antiquities and Museums regarding the management of the Marble Court area (see Chapter 4, this volume) and Bin Tepe. The Turkish Antiquities Law 1710, the *Sayılı Eski Eserler Kanunu*, was passed in 1973. It moved beyond "works of art" to consider terms associated with movable and immovable monuments as well as archaeological, historical, and natural site areas. In 1974, Sardis requested that Turkish authorities manage those areas that had been restored at Sardis (e.g., the Marble Court Complex,) as well as those outside of the current research strategy of the Expedition (i.e., Bin Tepe). Funds were insufficient within the Turkish General Directorate, and a request was made by the Turkish government to the Expedition to continue to manage not only the main excavation house and ongoing excavations, but also the Marble Court and Bin Tepe.[27]

Changes to Turkish national legislation followed with Law 2863 (Conservation of Cultural and Natural Properties), revised again in 1987 (Law 3386), which introduced Regional Councils for Conservation of Cultural and Natural Properties (*Yerel Koruma Kurulları*). The responsibilities of the Protection Councils consisted of site registration as well as the subsequent stages of informing the Directorate of Land Registry (*Tapu Müdürlüğü*) and registration for landowners of future changes. The details were defined under the March 4, 1988, decision of the Supreme Council for the Protection of Cultural and Natural Assets (*Kültür ve Tabiat Varlıkları Koruma Yüksek Kurulu*):

a) It is strictly forbidden to carry out agricultural activities that rely on plowing and to plant trees on the mounds and tumuli, which are designated as the 1st degree archaeological sites. Only already planted trees may remain and be harvested.

b) Every type of excavations (except those of scientific purpose) shall be prevented; soil or stones shall not be removed; and lime, stone, brick or marble quarries shall not be opened.

c) In times of necessary situations, it is required for public and private institutions to obtain a permit from the protection councils: to erect poles, to construct water channels and pipelines and to build highways, etc.

d) It is decided by the local administrative offices that [archaeological sites and tumuli] shall be marked by boundary stones or enclosed by metal wires, marked by descriptive signs by the Ministry of Culture and shall be provided with a security guard.[28]

In 1989 the Manisa Museum put forth two proposals for the management of Bin Tepe. The first proposed sixteen archaeological zones corresponding to clusters of tumuli. The second proposed complete, restrictive zoning (the highest level, *1. derece arkeolojik sit alanı*) for the entire area, except for the villages (designated with the lowest level of protection for archaeological areas, *3. derece arkeolojik sit alanı*). According to villagers, there was no consultation regarding boundaries (nor was it required under the legislation). To avoid "difficult village politics," the Manisa Museum recommended the highest level of comprehensive zoning for the entire area. Archaeologists familiar with this situation recall similar sentiments of frustration regarding negotiations between agriculture and national legislation for the protection of archaeological sites.

The path of least resistance at the time was to move forward with blanket protection. Decision 1.6.1.1990-1946 (1.1. 1990) remains the current legislation for Bin Tepe.[29] This decision forbids activities that alter the surface of archaeological sites and monuments, such as mining, quarrying, planting, and plowing. It stipulates a 30-meter buffer zone around each tumulus and a 100-meter buffer zone around each village or farmstead. In practice, most of these regulations were not put into place in the early 1990s, nor have they been enforced.

The situation was made more complicated when results from the CLAS (2005–2013) revealed archaeological sites located between the tumuli.[30] However, these could not (and still cannot) receive additional protection from the Izmir Second Protection Council because they are already subsumed within the original designation that covers the entire 72 square kilometers. This bureaucratic kink further complicates protection for the newly documented Upper Paleolithic, Early Bronze Age, Iron Age, Roman, and Ottoman era settlements because the original legislation was intended only for tumuli, and thus these subsurface sites are churned by powerful, deep-plow practices as well as the deep roots of fruit trees and vineyards.

In early 2013, at the request of the Ministry of Culture and Tourism via the Manisa Museum, a proposal was put forward to the World Heritage tentative list in Turkey, *World Heritage Nomination for Sardis, Bin Tepe, and the Cultural Landscapes of the Marmara Lake Basin*. This document discussed the natural and cultural assets of the area from past to present. In April 2013, *The Ancient City of Sardis and the Lydian Tumuli of Bin Tepe* (No. 5829) was posted. The proposal had been modified: the cultural and natural areas of the entire Marmara Lake Basin (i.e., far beyond Bin Tepe) and the living landscape component had been cut. Instead, the listing celebrated only the tumuli in the hinterlands of Sardis:

> The elite Lydian tumulus cemetery of Bin Tepe lies around the southern edge of the Marmara Lake. These tumuli are the most conspicuous ancient landmarks of Lydia, visible from afar and marking the region as a place of peculiar, haunting significance. Covering some 74 square kilometers, Bin Tepe is the largest tumulus cemetery in Turkey. Today about 115 tumuli survive in Bin Tepe.

This decision further vilified local communities who pursued agricultural development endorsed by both national regulations and international funding engines. Contrary to what many might presume with a nod to the potential for future World Heritage status, no changes have been implemented: farming continues apace without any concrete measures for wise-use. The Manisa Museum has coveted priority to Karnyarık Tepe, effectively shutting Sardis out. Ironically, however, this tumulus continues to be known locally as "Amerikan Tepe." Research-driven excavations (tumulus or archaeological site) in Bin Tepe are next to impossible given the hurdles of bureaucracy.

Unfortunately, in recent years looting on many tumuli has increased, likely a result of the economic downturn and the expansion of the illicit antiquities trade stemming from conflicts in Syria and Iraq.[31] What is more, the landscape continues to be modified by regional road networks. As of winter 2018 the expansion of the two-lane highway to four lanes through Bin Tepe is among the largest earthmoving projects since the construction of the water infrastructure and before that the tumuli themselves. The future course of a subsurface gas pipeline (proposed in the fall of 2017), too, will require a lot of earth moving in this sacred landscape. Not only is the cultural heritage at risk, but so, too, is the natural heritage, from the wetlands of Lake Marmara to the forests that ring the western and northern boundaries of the Basin.

Natural Heritage

In 1990 Turkey formed its first Ministry of Environment, yet policies regarding national environmental frameworks and harmonization with international standards have been slow to emerge, and the structure of this ministry has taken on various incarnations (Ministry of Environment and Forestry, 2003; Ministry of Forestry and Water, 2011; Ministry of Environment and Urban Planning, 2011). Turkey became party to the 1971 Convention on Wetlands, known as the Ramsar Convention, in November 1994. Despite Ramsar sites and IBAs designated by the IUCN located in Turkey, there have been considerable challenges to sustainable management of key zones.

At the 40th World Heritage meetings in Istanbul in 2016, the Turkish delegation pursued this agenda. An evening side event was held with the title "World Natural Heritage Potentials of Turkey and World Agricultural Heritages." The first part was organized by the Turkish National Commission for UNESCO, the General Directorate for Protection of Natural Assets, and the General Directorate of Nature Conservation and National Parks of Turkey. The IUCN, the Turkish National Commission for UNESCO, and the Ministry of Agriculture of Turkey organized the second half of the evening. Evidence that this session reflected a wider momentum within Turkey to effect change appeared in the Turkish tentative list for World Heritage: in 2016 the Kızılırmak Delta Wetland and Bird Sanctuary became the first site proposed under the natural category in Turkey.

Of the seventy-plus sites listed on the Turkish tentative World Heritage list as of Spring 2018, only a handful pertain to natural heritage, such as the Güllük Dağı-Termessos National Park (2000)[32] and Lake Tuz Special Environmental Protection Area (2013). The disjuncture in cultural and natural listing reflects the bifurcation in the politics of development and environmental conservation and epitomizes the challenges for the future of natural heritage. Such efforts are made more complicated by the relatively young governmental apparatus. Advocating for wise-use management (not even necessarily protection) of natural settings often requires a very long, arduous battle with bureaucracy.

From an ecological perspective, Lake Marmara is of primary importance to Turkey as the largest fresh water source to support the Gediz Delta Bird Sanctuary, a Ramsar site (1998 TR 945) and a Ramsar Hot Zone. The significance of this wetland zone pertains not only to western Turkey, but also to the entire Mediterranean region. The rich biodiversity of the Gediz Delta includes a series of lagoons, reed beds, shrub lands, saline pools, and agricultural zones. Of significance are the 80,000 species that breed each season, and notably of these 28 are endangered. In sum, both the Gediz Delta and the Lake Marmara Basin are among the most significant large-scale wetlands in western Turkey.[33]

The mid-twentieth-century modernization projects (see Chapter 3, this volume) shifted not only the ecology of the Gediz Delta and the Lake Marmara, but also the cyclical nature of the annual flooding and water volume of the Gediz and Alaşehir rivers. What is more, the scale of pollution from intensive agriculture and industrialization over the last three decades (e.g., Nif Valley and Menemen regions) has severely restricted the volume of clean water available for the lower Gediz River system and ultimately the Delta itself, making upstream water in the Marmara Lake Basin of critical importance.[34]

With the privatization of the State Hydraulic Works in 1993, and with related resolutions for the allocation of water and associated fees, the Ministry of Agriculture was designated to issue and monitor fishing permits for Lake Marmara. In response to the extreme droughts of 1993 and 1994 (i.e., the Lake dried up) new infrastructure was installed to pump water from the Ahmetli regulator/holding area back into Lake Marmara, cutting through low-lying areas of eastern Bin Tepe. Over the next decade resolutions to irrigation issues were tackled during two EU International Cooperation Programme projects: Sustainable Management of Scarce Resources in the Coastal Zone and Optimization for Sustainable Water Management, sponsored by the EU FP5 and FP6 programs, respectively. The process was participatory and had good results: mandates for water releases and irrigation requirements were followed.[35] Yet discussions from farmers in the last three years (2015–2018) reveal that corruption and illegal pumping have increased.

In addition, regulations and protocols of wetland management to sustain ecological habitats in Lake Marmara have never been fully realized. This despite two significant benchmarks: in 2008 the Ministry of the Environment and Forestry established protection zones for Lake Marmara, delineating a core wetland area, a larger ecological boundary, and a buffer zone for the watershed. The Lake gained further distinction in 2009 as an IBA (TR015), designated by Bird Life International with inclusion on the Red List of the IUCN. These have drawn attention to the Lake's endemic species of fish and a variety of birds.[36] Still, none of these international badges has led to enforcement of basic mitigation measures: abandoned irrigation pumps and dilapidated structures still dot the shoreline and illicit dumping of trash continues. The contrast between the old rusted-out pumps against the sleek, fire-engine-red new ones is stark, as shown in Figure 5.9. These now contrast with new signs installed in late summer of 2017 that indicate the updated status of the Lake as a national protected area.

Many of these issues could have been tackled as part of the preparation process for the World Heritage dossier (tentative list), especially if the Marmara Lake Basin had been considered as a mixed-use Cultural Landscape, as initially proposed. The rush to list Bin Tepe (and Sardis) took precedence over sustained discussions with various ministries for a mixed listing. This example demonstrates how archaeologists become entrapped by statecraft, and even act as primary agents for reproducing entrenched, bureaucratic systems. While aware of the less-than-desirable decisions, particularly the decision to exclude

FIGURE 5.9 Water pump on the edge of Lake Marmara, 2016. Courtesy of the author.

Marmara Lake, the archaeologists are also beholden to the state for access, for a permit. In this way, the archaeologists were obliged to support the World Heritage process. In sum, the situation in the Marmara Lake Basin reflects the limits of transparency, but also the state's ability to govern her subjects effectively.

Summary

This section has provided an overview of the historical and current archaeological and natural heritage zones in the region. In the following sections, I trace parallel agricultural policies. The privatization of land, access to assistance programs, and infrastructure (road and irrigation) encouraged the expansion of agriculture, and in so doing these mechanisms bolstered individual claims to self-identity, in direct contrast to heritage policies that aimed to protect the natural and cultural landscapes under the pretext of public appreciation and access.

Land

As discussed earlier in this chapter, in the Marmara Lake Basin at least two Yörük tribes, the Tekeli and Karayahşi, practiced annual migrations until the late 1940s. With the national mandate to settle, they did so, and today live in the villages of Tekelioğlu and Karayashi.[37] Formal deeds for homes or lands, however, were not issued, and the area was not given priority for mapping until the early 1950s. This followed the 1950 *Gediz Basin Study* prepared by Knappen, Tippetts, Abbett Engineering Company of New York (KTA), which is explored in detail in Chapter 4 (this volume). It included design plans for the hydroelectric dam, canals, regulators, and reservoirs as well as a general assessment of the landscape (soils, water flow, etc.) and the projections of future demand for electricity and water in Izmir. The report also stressed the untapped potential in the countryside.

The implementation of the Gediz basin hydroelectric project shifted priorities in the Marmara Lake Basin and abutting areas. Maps were essential for the implementation of the waterworks and the requisite massive manipulation of the landscape. In some cases, entire towns were redesigned. For example, in Adala mid-twentieth-century modernist principals for public gardens and parks were incorporated into the infrastructure of canals, regulators, and reservoirs. The same was true at the residential structures in the plain, those that provided a network with the respective pump houses. They, too, had extensive residential areas with manicured gardens. Yet, the most elaborate gardens were those of the company compound near the hydroelectric station, dam, Demirköprü reservoir, and spillway.[38] Today the once-elaborate gardens have been let go; only the planting borders and ruined stone walkways indicate their former glory.

Without annual inundations, the expansion of villages and large areas of farmland followed, creating entirely new social and economic systems on the face of the plain.[39] Changes to the landscapes of Bin Tepe, in contrast, were less extensive. The high elevation of the limestone ridge meant that this area was naturally above the annual flood line, and therefore the face of the landscape was not canalized, but large diversion canals were dug along the eastern and southern low-lying boundaries of Bin Tepe to link Lake Marmara and the Ahmetli regulators. At each of these points, residential compounds were built.[40]

Regardless of whether lands were slated for inundation, infrastructure, or no transformation at all, the cadastral mapping and formal issuing of deeds proceeded throughout the region. If lands were located outside zones required for infrastructure, deeds for private property were still issued. This was the case with most land in Bin Tepe. As mentioned previously, elders in the village of Tekelioğlu credit their acquisition of land to US assistance. Indeed, the policy shifted individual sovereignty and long-term prosperity for many people in the Gediz Valley.

In contrast, if lands were located in areas where infrastructure was required, compulsory expropriation was implemented, often immediately following deed issue (i.e., the same day).[41] Along the western shores of Lake Marmara (i.e., the community of Hacıveliler) expropriation divested farmers of lands promised under the 1936 and 1945 legislative initiatives.[42] However, annual inundation has been less than expected, and thus lands without title continue to be farmed. In contrast, farmers in villages of Bin Tepe (Kargın, Kestelli, Kenderlik, and Tekelioğlu) who had, just a decade before, followed a transhumant lifestyle and/or who had been subject to the feudal-like *ayan* system of the Karaosmanoğlu, overnight became landowners.

As villagers acquired land, the power of the wealthy Ottoman-elite dwindled, and in turn these families and individuals looked for new ways of expressing (and clinging to) power. As discussed by Bouquet and highlighted in Chapter 4 (this volume), it was common for members of the pre-1923 powerful *ayan* families to use connections and experience to enter politics such that they could steer legislative initiatives as much as possible in their favor. This was the case with Mr. Fevzi Lutfi Karaosmanoğlu and the lands attached to the Bağırgan estate in Bin Tepe.

Neoliberal Climate

By the mid-1950s, and especially under President Adnan Menderes, Turkey began to implement new policies to encourage market competition. These changes allowed farmers to tap loan packages for tractors. The first tractor for a family (or often collectively owned by a village) was a point of great pride, as shown in Figure 5.10. During this time, Turkey became a rising competitor in the world grain market, and agricultural engineers sought to expand this sector.[43] Recall that The first full five-year plan was implemented between 1963 and 1967.[44] This corresponded to the 1963 Turkish Seed Law No. 308, when the Ministry of Agriculture began to standardize seed production and trade.[45]

Foreign investment, too, sought collaboration for research and access to knowledge and new technologies. Turkey was a founding member of the 1961 Organisation for Economic Co-operation and Development (OECD). Shortly after, in 1963, Turkey applied for Associate Membership in the European Economic Community (EEC), known as the Ankara Agreement (implemented in December 1964). Farmers relied on government subsidies and set prices for specific commodities through government purchases and sales. The State Economic Enterprises, the Agricultural Sales Cooperative

FIGURE 5.10 First tractor, 1961. Courtesy of the Şener family.

Unions, and the state-owned banks were all part of the structure that facilitated this network. The 1980 coup d'état dampened relationships with the EEC. It was not until 1986 that negotiations under the Ankara Agreement were again on the table for Turkish farmers.

Under the Maastricht Treaty in 1992, which streamlined mechanisms for the European Union, relationships shifted yet again. The broadening of Turkish participation in major markets led to closer ties with international financial institutions, including the IMF, and the World Bank. These partnerships also facilitated a viable structure for TNCs and neoliberal models.[46] In the late 1990s the World Trade Organization promoted liberalization of policies for the trade of agricultural commodities. Predictably, large firms in the United States, Canada, and the EU looked to new markets and viable areas for expansion.

Neoliberal agendas and policies endorsed increased efficiency and high-value crops. Aydin argues that in 1989 "full capital account liberalization" jump-started a "financial revolution" in Turkey and a move toward dependency on foreign firms (especially their capital). He explains that this structure was "intended to ensure savings and investments, [but] instead led to speculative foreign capital flows . . . which culminated in the collapse of the financial system and the emergence of a severe economic crisis in 1994."[47] In 2001 Turkey signed an agreement with the World Bank for implementation of the Agricultural Reform Implementation Project. As Demirdöğen and colleagues explain, "within the scope of this project, state-owned enterprises were privatized, the labor forces decreased in the remaining SOEs, farmers were registered with the Farmer Registration System."[48] In addition, parallel 2000 IMF policies favored the transformation of agrarian structures and expansion of TNCs as well as initiatives to align Turkish policies closer to those of the EU's Common Agricultural Policy.

Mexican Wheat and Internationalism

In light of the changes in agricultural policy in Turkey from 1960 onward, I now take a step back in time, to the mid-1960s. I examine the impact of these trends in Bin Tepe. The 1960s marked the period when Turkey began collaboration with USAID. One example is the 1967 USAID-Turkey Mexican Wheat program:

> Eight members of the American Team were Oregon and Washington County Agents and four were leading farmers from these states. The Team arrived

> in Ankara on August 20. They were given a five day briefing on Turkish agriculture and customs, the Mexican wheat program including the implementation plan and the role of American technicians. Following this briefing they were assigned to their posts.[49]

Charles Hindes was the supervisor for the Aegean region, and Gus W. Hokanson, a country agent from Washington D.C. (i.e., not a farmer), was assigned to monitor operations in Manisa, which included over 10,000 hectares (approximately 25,000 acres). The Mexican seed arrived at the port in Izmir between July and August 1967 and was planted shortly thereafter. One-day trainings took place during impromptu meetings in teahouses, where brochures and guidebooks were also distributed. Humphrey's 1969 report claims that there was "a devotion to a cause probably never before experienced by the Turkish extension workers."[50] Supposedly, more than 60,000 Turkish farmers took part in the programs.

Farmers were told to shift from a broad sowing pattern that used a lot of seed to grain drills, effectively using much less seed. In addition, after this process they had to apply a heavy coating of fertilizer—which had never been done before.[51] From the US perspective, the program was a success:

> There can be no doubt that the boldness and commitment of the Minister of Agriculture in Turkey to a wheat revolution was decisive. He shocked his own advisers in calling for, and obtaining, seed and fertilizer for 170,000 hectares the first year and 650,000 hectares (1,605,500 acres) the second in their accelerated program. In the affected areas, a threefold increase over the amounts grown with native varieties of seeds and with native agricultural practices was achieved.[52]

That the program brought together US and Turkish colleagues fulfilled the educational exchange component of USAID's goal, and it delivered immediate returns on investment.

Elders who recall the Mexican Wheat program in Bin Tepe remember groups of foreign men (i.e., trainers) who came to show local farmers what to do, despite their (the farmers') decades of experience. Turkish farmers doubted the future success of Mexican Wheat given the soil conditions in Bin Tepe, which in fact were not ideal. The report for the region of Manisa, and specifically areas near Salihli, corroborates their accounts. The challenges with the poor soil conditions were also exacerbated by a period of drought. The solution, according to Hindes' November 18, 1967, report, was subsoiling—deep tillage of the landscape's surface.[53]

Despite their doubts, Bin Tepe farmers went along with the program with an eye toward long-term benefits: they expected the wheat to disappoint. In return, however, they predicted that "the foreigners" would leave, but the infrastructure (i.e., paved roads and enhancement of the irrigation systems) and equipment from extremely favorable loan programs (i.e., tractors and plows) would stay. The gamble worked: the crop failed in Bin Tepe and USAID pulled the program from the area. Yet the tractors remained with the farmers, who were able to secure loans for more equipment to proceed with subsoiling techniques, which were successful with local wheat varieties as well as other crops.[54]

While USAID did not have success in Manisa, the region remained on the international radar through connections to the World Bank. A member of the Karaosmanoğlu family from Manisa, whose name was Attila, took an interest in agriculture.[55] He assisted in the policies and implementation of the five-year plan (1963–1967). He also contributed to Turkish harmonization with the OECD.[56] In 1966 he took a position with the World Bank. In 1971 he returned to Turkey for a short period and served as Deputy Prime Minister. His vision focused on people and solutions to poverty:

> I always believed that governments need to be concerned about the welfare of the poor, that governments are responsible to ensure the equality of opportunities, that governments have particular responsibilities in education and basic health. For instance, when I was in Planning we carried out surveys in 40,000 villages in Turkey to determine their priorities. Did they need roads? Education? Health care? Where did they want to start? Some of the old bureaucrats thought we were crazy because we would increase the expectations of people. Yet for me increasing their expectations and desires was in fact something that needed to be done.[57]

In 1973 Attila returned to the World Bank where he worked until 1994. Over his tenure, he held many positions that engaged with Europe, the Middle East, and North Africa, and eventually with Asia.[58] He worked closely with Robert S. McNamara (President, 1968–1981) and Ian Peter M. Cargill (senior vice-president, 1974–1980). Both McNamara and Cargill promoted both the modernization and capitalization of agriculture, ushering in the neoliberalism of the 1980s. These ideals were not, however, compatible with the long-term viability of the small family-farm, nor with the Kemalist vision of the village as a microcosm of the nation. The shift to corporatism for family-run agribusinesses in Europe (i.e., expansion of the family farm) came to represent the norm rather than the exception in development.[59] Such a model would have upset the pace of life in Bin Tepe considerably.

Organic Agriculture

The trajectory in Bin Tepe did not follow the neoliberal model, owing to the involvement of the German company of Rapunzel Naturkost and their emphasis on organic agriculture and cooperative village programs. Scholars trace the European organic movement to Rudolf Steiner's 1924 vision, followed by supporters Eve Balfour and Albert Howard in Britain, and Masanobu Fukuoka in Japan.[60] In 1972 the non-profit International Federation of Organic Agriculture Movements was established. Shortly thereafter, in 1974, Joseph Wilhelm founded Rapunzel in Germany. He made several visits to Manisa and specifically the region of Salihli throughout the mid-1980s to explore controlled organic cultivation. In 1989 Wilhelm established the village project in Tekelioğlu, one of three villages located in Bin Tepe.

The evolution of the organic agricultural movement in Turkey paralleled Rapunzel's investment in this sector. In 1992 the non-profit Turkish Association of Organic Agriculture Movements was formed, followed in 1994–1995 by regulations of the Turkish Ministry of Agriculture and Rural Affairs for plant and animal production with organic methods. During this period, EU policies dictated mechanisms for trade of organic produce (i.e., EEC No. 2092/91, implemented in 1993 for organic agriculture, and EEC 1804/99, implemented in 2000 for organic products from livestock). The Turkish "Organic Farming Law" went into place in December of 2004.[61] Until recently, organic agriculture accounted for just under 2 percent of agricultural production in Turkey and remained primarily vested in small, family farms. Oversight of and training for organic agriculture in Turkey was primarily organized via contract farming, and the majority of parent companies were those with a keen interest in export, including Rapunzel.[62]

Initially Rapunzel concentrated on dried fruits (sultanas, figs, apricots), hazelnuts,[63] and the cultivation and production of organic cotton fibers.[64] Rapunzel supplied the seed, fertilizer, and knowledge; farmers supplied the land and machinery (both, in retrospect, courtesy in large part of US assistance). In this way, Rapunzel built on the earlier infrastructure of the various Turkish and US assistance programs. Even with a model driven by external markets that fostered a patron-client relationship through contract farming, villagers in general saw Rapunzel in a favorable light because of the company's commitment to an entire community, which had little comparison to the typical corporate engines of international agro-companies or the legacy programs of the Marshall Plan and USAID, which promoted individual entrepreneurs. Recall that through the US programs, only individuals, not cooperatives, were eligible for loans.

Private conversations and practices of daily life, however, reveal the vexed relationships, not only with Rapunzel, but also with the ideal and purity of "being organic." The elders of Tekelioğlu grasp the irony of the elite-driven organic markets for those who've never farmed. To many, the urban promotion campaigns of living "pure" are absurd. They describe with dry humor (and often a smirk) their "pure" Yörük roots. Their narratives envision their ancestors living in harmony with the land. To be sure, many praise the benefits of organic methods and eating organic. They claim that their lives are less stressful than those of large-scale, corporate farmers, and they believe that their farming methods will help them avoid cancer. But, to combat household pests, a liberal coat of insecticide (such as Raid®) is often sprayed in homes; and bathrooms, too, are cleaned with non-organic cleaning supplies.

Farmers also find Rapunzel's requirements exhausting. Older women perceive the required daily logs, referred to as the "jail book," with apprehension and distrust. What the local community hasn't fully understood is that these requirements are imposed not necessarily by the firm (Rapunzel) but by the bureaucracies embedded in EU and Turkish regulatory policies for organic certification. That is, the spotlight isn't specific to them, nor to their village and lands. On a more duplicitous level, however, in hushed conversations with community members, farmers, and DSI personnel, people spoke of "nocturnal nacros." These rumors of farmers in the region who spray non-organic pesticides in the deep of the night also become fodder for disgruntled neighbors, who threaten to report (or even fabricate) organic misdeeds to avenge personal feuds.

The situation becomes more complicated once we consider the impact of Rapunzel's relationships on the financial position of farmers. Rapunzel-villages and Rapunzel-farmers are seen in favorable light by banks because Rapunzel acts as a pro bono/de facto gatekeeper, and thus mitigates risk. For the landscapes and villages in Bin Tepe there is another layer of complexity. Banks usually avoid loans for agricultural activities to be carried out on plots of land in registered archaeological zones, but exceptions have been made for "Rapunzel lands" in Bin Tepe. This foreign partnership with a high-profile "sponsor" from an EU country such as Germany generates not only economic remuneration, but also conditions that sanction deliberate violation of Turkish national patrimony laws. Through their tacit approval for agriculture and their inaction to reach suitable solutions companies have been complicit in twenty-five-plus years of ongoing violations whereby bank loans are issued for agricultural practices in archaeological areas.

Legal Landscapes? The Empowered and Entrepreneurial Farmer

In the preceding sections, I have described the history of archaeological zoning, archaeological excavation, and survey programs as well as the listing of Bin Tepe and Sardis on the 2013 World Heritage tentative list for Turkey. I've also presented the development agendas for formal mapping and deed registration, evolving agricultural initiatives, and the ecological zoning for Lake Marmara. While heritage is supposedly highly valued, the agricultural lobby (foreign and national) presents a formidable hurdle. For the archaeological landscapes of Bin Tepe, programs and policies aiming to phase out tobacco and to support high-return organic crops further entangle this debate.

This situation is a result of over sixty years of development agendas failing to consider a holistic, sustainable approach. A review of how this situation arose is necessary: in 1954 the US Marshall Plan spurred Turkey to issue deeds of ownership, which paralleled Turkish policies of vesting sovereignty with individuals. Farmers in Bin Tepe benefited. Favorable subsidies to secure machinery in the late 1960s were made possible through Turkish and US programs, including USAID's Mexican Wheat project. The neo-liberal policies of the 1980s further galvanized foreign interest in agricultural investment in Turkey. Rather than embrace a full neo-liberal approach, the community of Tekelioğlu partnered with Rapunzel.

The turn to olives after 2000, however, shifted the Rapunzel arrangement considerably. There are limitations for selling Turkish olive oil within EU borders because the EU olive industry is accountable to Italy, Spain, and Greece—all EU members (unlike Turkey). Rapunzel had supported some olive-tree production for oil to pack organic sun-dried tomatoes, but not on the scale of primary production. In the early 2000s, those farmers who had done exceptionally well working with Rapunzel began to pivot out of a full-contract relationship to join other independent entrepreneurs investing in olive orchards in Bin Tepe. Adding fuel to this growing fire, several new investors to the region purchased land in Bin Tepe for olive groves, unaware of the strict archaeological zoning. Once informed of their violations, they pursued litigation. Their cases demonstrated a lack of transparency and due process in setting archaeological boundaries as well as communication with the wider public and potential investors from outside the area.

Arguments were based on failure of the Manisa Museum and the Second Protection Council in Izmir to notify stakeholders and to make such information publically and easily available for new investors. They argued that tumuli should remain single, standing monuments with the highest (first degree) archaeological protection, yet zoning for areas in between could be lowered (third degree) if no archaeological material was present. In addition, farmers expected fair compensation for expropriated property.

This decision resulted from reviewing questions of equity extending from regulations and legal cases in the mid-1990s that farmers could exchange private lands in archaeological zones for public lands (which would then be held privately).[65] Farmers argued that the quality of lands offered for exchange was not comparable. The soil conditions were often not as rich and/or the plots were too distant from established infrastructure. Details of this process had been given in announcements made over loudspeaker systems connected to village minarets. For those who were interested in exploring the possibility, they had to gather, compile, and bring all files (including those of the lands they wished to acquire as part of the exchange) to a regional directorate, or directly to a branch of the then Ministry of Culture—all within sixty days. The exchange would have been determined by value of the land only; that is, value added (i.e., vineyards and/or orchards) was not considered. Details pertaining to adjudication of multiple requests for the same public property were murky at best. Unsatisfied, farmers persisted, eventually bringing cases to court.

The courts disagreed with the Izmir Second Protection Council's decision and ruled that full compensation for Bin Tepe lands must be paid.[66] Three other decisions pertained to a failure of due process in 1954 and 1961 whereby the possessor of the land (i.e., one claiming use of more than twenty years) had not been properly deeded the parcel.[67] Regulations of archaeological zoning failed to recognize and rectify ownership, which the courts ordered. Farmers continued to press the issue with the municipality and the governor's office in Manisa. The Ministry of Agriculture advocated on their behalf, also taking issue with what they saw as prohibitively strict regulations that interfered with economically advantageous crops. Having been handed their heads back to them on a number of occasions, the Protection Council and the Manisa Museum backed away.

The coalescence of events between 2004 and 2008 increased the complexity of the situation in Bin Tepe. With rulings in their favor, and the newly implemented 2004 Organic Farming Law, farmers moved forward with establishing organic olive orchards in Bin Tepe, a process they argued offered greater variegation to an otherwise desolate, barren landscape, especially during the dry summer months, when the leaves of olive trees are green. In 2007 a local company named Bintepe Olives, located in the adjacent Büyükbelen valley, placed twelfth in the World International Organic Olive Oil contest held in Italy. Further solidifying the growth of this sector was the first Izmir Organic International Fair in 2009. This was followed in 2011 with the first Izmir Olive Oil and Technology Fair, *Olivetech*. And, in 2012, another local company, Alyattes Organics (with extensive fields in Bin Tepe), took second place at the International Extra Virgin Olive Oil Competition held in Japan.[68] With each new positive outcome organic agriculture for olives in Bin Tepe expands, which also parallels this growing sector in Turkey.[69]

From the perspective of the Ministry of Agriculture and Rural Affairs, the profit margins are welcomed and encouraged.[70] Yet, without a commitment to the harmonization of conflicting policies and long-term solutions (such as investment in GPS guidance systems for tractors so that archaeological areas could be avoided and/or fair market prices for expropriation), the region remains in legal limbo.[71] Regulations pertaining to the cultivation of the landscape have been extremely lax, and with each new agricultural season more powerful farm equipment, as illustrated in Figure 5.11, is procured by farmers.

The flexibility for pursuing agriculture, however, has not been true for construction: regulations prohibiting new buildings in archaeological zones remain strictly enforced. For example, in the spring of 2011 the Büyükbelen municipality proposed to build an olive oil production plant just inside Bin Tepe's western boundary, which was included in the highest degree of archaeological zoning. The action was thwarted by the Manisa Museum. Yet, the property of the former office and depot for Alyattes Olives along the southern shore of Lake Marmara (also in a first-degree archaeological area) was allowed to remain, largely owing to the liminal period of when the facility was built. What is more, within the third-degree boundaries of Tekelioğlu, farmers have expanded their footprints considerably, ignoring regulations or willingly paying penalties

FIGURE 5.11 Agricultural equipment in Bin Tepe, 2011. Courtesy of the author.

(and thus grandfathering their operations). As shown in Figure 5.12, agricultural fields abut the largest man-made monument on the Anatolian peninsula, the tumulus of Alyattes.

Discussion

I argue in Chapter 3 that the designed landscape of water management and agricultural development has created a dependency relationship between city and countryside (i.e., Izmir and her hinterland), a concept that shaped Le Corbusier's original thinking and the policies endorsed under Soviet and later (and much more extensive) US assistance programs. Half a century later, the original intention continues to be realized in the middle Gediz Valley. Due to lack of oversight, however, the tumuli in Bin Tepe and the wetlands of the Marmara Lake Basin have been subject to arbitrary decisions regarding highway expansion, plunder, and organic agriculture. The primary driver of the organic agriculture is urban centers. Since the 1950s, not only has Izmir expanded (it is now Turkey's third largest city after Istanbul and Ankara), but so, too, has the province of Manisa, now registered as a metropolitan area (i.e., *Büyükşehir*).

In assisting with the preparation of files for the World Heritage tentative list, an attempt was made to stress the living landscape nature of the

FIGURE 5.12 Organic olive groves and other agriculture abutting the tumulus of Alyattes, Bin Tepe. Farmer on tractor is seen in the foreground 2017. Courtesy of author.

Marmara Lake Basin as well as the mixed natural-cultural value, from past to present. The goal was to look before and beyond the Iron Age tumulus landscape of Bin Tepe. The reasoning was based on an understanding of how this landscape works: local pride and daily behaviors in Bin Tepe reside in the firm conviction of the value of the rural, family farm to Turkey; and local sovereignty in the middle Gediz is linked equally to the nation, development dollars, and private enterprise. Over time there has been—and there continues to be—a sustained momentum of significant social mobility and economic return to individual families, marked by personal material benefits, expansion of infrastructure for farms, and access to education. These changes are tied to global modes of behavior pertaining to the perceptions of organic produce and by national, regional, and international networks vying for connectivity into, within, and beyond Turkey. The result is that archaeological heritage is not a priority.

As Sardis Director Hanfmann predicted in the late 1960s, the initial investments in tractors, rural roads, and water management have transformed the landscapes of Bin Tepe in ways that have been detrimental to its archaeological heritage. But these changes have had a significant, positive impact on the spirit of the entrepreneur, the farmer. The very US engine that had been poised to make a significant difference to restoration programs and regional management of archaeological resources in concert with local communities crumbled in the late 1960s. Yet, in its place was another US engine, that of mechanized agriculture and water diplomacy, which capitalized on Ottoman-era elites seeking to carve out new positions of power and prestige, and on communities caught in the arc of national agendas aimed at large-scale management of rural territories. Turkey ushered in programs with a goal toward "betterment" of the national agenda, which in turn provided security for US investments. The goal, then, was to harness the citizenry such that they would not only accept the changes but also become the primary proponents for these programs.

Ever since, tumuli and archaeology have been compromised by development, whether tractors and/or the infrastructure required for roads, irrigation, and electricity. Today Koca Mutaf Tepe (Aylattes), Karnyarık ("American Tepe"), and Kırmutaf Tepe still command the landscape, yet the steel lattice electric transmission towers, as shown in Figures 5.11 and 5.13, remind us of the monumentality of contemporary and future

FIGURE 5.13 Electric transmission towers carry power from the Demirköprü dam through Bin Tepe toward Izmir, 2017. The towers dwarf tumuli in Bin Tepe, including the second largest, Karnıyarık, also known as "American Tepe." Courtesy of the author.

FIGURE 5.14 Scarecrow protects tobacco fields (foreground) in Bin Tepe. In the background are olive groves and tumuli, 2017. Courtesy of the author.

infrastructure projects. What many fail to understand is that farmers and developers act in response to the national and international policies that favor their positions, not necessarily against cultural heritage. To be sure, however, they do not advocate for preservation either. The unfortunate consequence is that protection of tumuli and those sites located between them has not been a high priority. A photograph (Figures 5.14) taken in June of 2017 captures the irony of the situation: a farmer placed a scarecrow to ward off birds from damaging crops, yet the nearby tumuli lack signage and other indications of their value and their buffer zones.

This case study demonstrates the challenges of preservation in rural agricultural areas where significant private property and public assets overlap. What we will never know is how Hanfmann may have been able to do just this had he been able to realize his 1965 proposal to the US Department of State for an environmental study and archaeological survey of Bin Tepe and the Gygaean (Marmara) Lake and a regional management plan. Of course, much would have happened in the intervening years, yet a solid foundation is often just enough to tip the scales, as the legacy of generously supported development diplomacy for roads, water, electricity, and agriculture confirms.

Reflections

THE NARRATIVES OF heritage in western Turkey will continue to unfold, with thousands more remaining untold. A conclusion for this book, then, is impossible, but a reflection offers a place to review the primary themes and provide additional analysis. The performance of heritage pageantry in Izmir and the Gediz Valley offers a spectacle that is a contemporary interpretation of events that have altered the landscape for generations. As such, an investigation of development projects has allowed for an exploration of how people have lived—and continue to live—among new installations, whether newly revealed temples, large-scale irrigation systems, urban parks, or resurrected Roman-era monuments. In these ways, the heritage project itself over the last century presents punctuated periods of development schemes that are as much reflections of Turkish policies as they are of international intervention and the rise of global corporatism.

With each new approach to development, we can trace the legacies of earlier programs with an eye toward the accelerated pace at which change occurs. Contradictions in policies over time are responsible for how heritage may be erased, forgotten, hidden, or celebrated. For heritage that confronts the status quo, there will be tensions. Conflicts are most raw when identity is challenged or when eradication (i.e., destruction of a place, enforced migration of a group) becomes a reality. These processes both create and enable feelings of mistrust, which in turn drive suspicions and distance efforts aimed at transparency and fluid dialogue. Yet, it is precisely key events that propel processes of change, which in turn provide opportunities for debate.

On the sunny, warm spring afternoon of March 21, 2017, I planned to take the afternoon off to walk from the Ahmet Piriştina City Archives in central

Izmir toward Cyprus Martyrs' Street in Alsancak, known for its friendly atmosphere and coffee shops. I never made it. Once I passed Canonica's 1932 statue of Atatürk on his horse pointing towards the Mediterranean (see Introduction, this volume),[1] I had planned to turn from the Republican Square toward the open Kordon and walk along the sea. Security fences blocked the entry, as they did all side streets for the next kilometer or so. As I approached Cyprus Street the area was clogged with people heading toward the Kordon through the one open access point. Armored police buses, too, arrived. Men in bulletproof vests, shields, and guns disembarked. A TOMA (armored vehicle with a mounted water gun) lingered around the corner. People were gathering for Nowruz, the Persian New Year, and the celebration of spring. Kurdish groups from Izmir and the region (and many from Iran, so locals told me) were making their way toward the Kordon for evening events, and the local Izmir policy were in place should things have become unsettled (they didn't).

Realizing that I wasn't going to be able to meander slowly along, I ducked into a café, and over an afternoon tea I reflected on the layered histories of this space. For the café owner (from Izmir), it was merely another event that required security and thus made his day more challenging than he would have preferred. For him, the mix of security and celebration was a new normal that he found uncomfortable yet not impossible to navigate. He said, "This is Turkey. This is Izmir! We are used to it. We survive." He was referring to patterns of both endurance and resilience during cycles of change. He was also clear that the Kordon increasingly represents a strategic flashpoint on Izmir's political landscape.

In the 1990s the Kordon had been slated for development. A seaside highway was in the works. People protested. They preferred a green, open park for pedestrians. Today at the end of the Kordon an abandoned concrete viaduct stands as a reminder that urban developments aren't always realized.[2] As the largest public space in Izmir outside the Kültürpark (see Chapter 3, this volume), the Kordon is not only a place for leisure but also a place for protest, a place to gather attention, whether on the 9th of September or other days of significance to specific groups, such as Nowruz. In this way, aborting the highway project sowed the seeds for public opinion to be heard as well as seen.

To be sure, however, the future holds drastic change for the transportation networks of not only Izmir but also western Turkey. Izmir will become increasingly connected to her countryside to the north, east, and south. Among the largest projects is the planned Izmir Bay Crossing, a joint project between the Ministry of Transport and the General Directorate of

Highways. While it leaves the Kordon untouched (as well as the abandoned viaduct), it alters Izmir's connectivity patterns in profound ways. Inciraltı/Güzelyalı and Karşıyaka/Çiğli will be connected via a 7.6-kilometer corridor that will reduce travel time circumnavigating the bay from 1 hour to 10 minutes. Plans for the Izmir Bay Immersed Tube Tunnel include a suspension bridge, an immersed tunnel, and an artificial island. The design plans for the project remind us of the mega-projects transforming Istanbul and Dubai.[3] In fact, many super transportation projects are altering the landscapes of Turkey.

Not since the Truman Doctrine of 1947, as discussed in Chapter 3 (this volume), has there been such unified, countrywide investment. As discussed in this book, development dollars then sought to link agricultural fields to markets with paved, village roads. Today highways serve as boundary breakers, cultural corridors, and economic engines that link cities to cities and countries to countries. Transportation plans and implementation paradigms, however, are no longer infused with the social doctrines of US Cold War containment policies. Over the last two decades Turkish businesses have shown steady expansion in Turkey and abroad.[4] The Foreign Economic Relations Board of the International Technical Consultancy Business Council of Turkey was formed in 2013 with the goal of expanding the Turkish international footprint in this sector, especially through Public-Private Partnerships (PPPs). In the last five years, presentations have been made to potential investors in Turkish, Arabic, English, French, German, Russian, and Serbian, reflecting key target areas for Turkish foreign policy in the Balkans, northern Africa, the Middle East, Russia, and Central Asia. The projects featured include the massive Third Bridge in Istanbul, the Gebze-Izmit Bay Bridge, the Nizzibi Bridge in Diyarbakır, and the Izmir Bay Crossing project. This spider web of infrastructure across Turkey means that cities and their hinterlands will likely see massive changes in the next decade.

Such transformations are underway in Izmir's countryside. Colossal earthwork platforms and massive concrete viaducts for a new highway system connecting Istanbul and Izmir rise from the valley floors on the outskirts of Akhisar and Manisa. New tunnels bore through the mountains of Belkahve and Sabuncubeli. This growing concrete footprint is hungry for electricity and water. Yet, the Demirköprü Dam and Lake Marmara reservoir can't meet the demand, nor can the new Gördes Dam. To supplement hydroelectric power, alternative technologies are quickly becoming necessary. Hillsides near Adala and Gölmarmara glimmer with photovoltaic panels, and mountaintops hover in the shadows of towering wind turbines. Capitalist in nature, these projects intend to promote PPPs through enhanced rural electrification.[5]

Private investment may be driving the engine of transformation that seeks to connect Turkey to her neighbors, but at the same time these new monuments of change are infused with collective memories of Turkish nationalism, all the more poignant the closer we get to the 2023 centennial of the Republic. While the use of monuments to signal political positions isn't new, the vigor and strategic placement of Atatürk's image in western Turkey continue to demonstrate opportune platforms for various constituencies during periods of charged politics.[6] For example, in the wake of national discontent, Tankut Öktem's 16-meter high monument of Atatürk gazing from Belkahve over Smryna, erected in 1991, celebrated the location of his final tea prior to Turkish forces entering the city on September 9, 1922. At the time this was the tallest Atatürk monument in Turkey. Today it still abuts the major highway, which now goes under rather than over Belkahve (the tunnel opened in the spring of 2017).

We need not look far to find another example in a more contemporary context. The 68-meter high Nationalist Forces Monument, also by Öktem, towers over the new highway as one departs Manisa toward the Sabuncubeli pass and beyond to Izmir. Planning here has been holistic. As routes shift, drivers will still pass this monument, whether driving over the pass or under it through the new tunnel opened in June 2018. Izmir's Governor Erol Ayyıldız and Turkish Prime Minister Binali Yıldırım attended the connection of the two ends of the tunnel, which took place at midnight on September 8, 2017, Manisa's liberation day from Greek forces in 1922.

Further into the hinterland this pattern of monumentality and commemoration of new infrastructure at strategic historical points persists. A golden (i.e., Lydian gold) west-facing ensemble of Atatürk on his horse flanked by two Lydian lions along the new ring-road around Sardis recalls the initial boundary line between Turkish and Greek forces following Article 66 of the Treaty of Sèvres, when the boundary fence ran up the Acropolis (see Chapter 2, this volume). These various monuments and others at key locations in the Gediz Valley confirm that the Turkish War of Independence continues to pervade the tapestry of a collective consciousness in western Turkey.

Assembling the future most often entails a re-reading of the past. I've argued throughout this study that the face of the landscape in the middle Gediz Valley is a proxy for the materialities and temporalities of the turn of the twentieth century through the first two decades of the twenty-first century. Indeed, many with an attachment to archaeological heritage in this region have been and continue to be (for now) foreigners with a strong American presence in the Gediz (Sardis), Austrian in the Cayster (Ephesus), and German in the Bakırçay (Pergamon). Political frustrations, however, indicate strain and the potential re-ordering of powers and permits.[7] In the case of Ephesus, Austrian cultural capital has weakened, while proposed concepts by Turkish developers have strengthened, including a plan to re-order the natural environment and literally bring the sea back to the port of Ephesus.[8]

In contrast to the unwavering of foreign concessions to archaeology (at least until recently), evolving narratives of transportation, extraction, and agriculture in Izmir's back yard have resulted in a mosaic of heritage futures and shifting international players. The French and British staked early claims to railroad concessions, and the Americans attempted to do the same with antiquities and archaeological sites (see Chapter 2, this volume). I've shown that through economic assistance plans, the Soviets moved into place as a strategic partner for rural development during the interwar period, only to be edged out by the Americans post–WWII. These performances were especially visible at the International Izmir Fairs at the Kültürpark and across territory in the province of Manisa. As of yet the heritage of these mid-twentieth-century landscapes remains hidden by the pressures of past mastering. With modernist heritage increasingly in vogue for World Heritage listings, however, we can anticipate the future heritage of these installations, especially as their capacity is replaced by alternative energies. Solar and wind-energy production uses the existing electric infrastructure that defines much of Bin Tepe's skyline (see Chapter 5, this volume), and soon a subsurface natural gas pipeline, too, will become part of this landscape's future.[9]

The capacity for the Turkish War of Independence, archaeology, modernism, and extraction to structure the human experience in the Gediz Valley pushes us to consider the relationships between not only political and social actors, as I've argued in this study, but also environmental issues, about which I have only skimmed the surface. While TVA bellowed its conservation focus in the United States, in reality it unleashed a catalyst that has altered entire river basins and people's lives from rural Tennessee to India—and not necessarily for the better.[10] When will the binary of nature–culture in heritage studies move governments, as well as the private sectors, to take far more seriously the ethical questions surrounding the holistic destruction of both natural and cultural heritage? Meskell has demonstrated obligations that we face not only to ourselves, but also to the environment (flora and fauna specifically) and the rhythms of a place.[11] Harrison, too, calls on how we prepare for "uncertain futures" in the Anthropocene.[12] Some may question doomsday scenarios, but many more ask how we make sense of transformative and largely irreversible modern-day monuments given our future trajectories.[13] How will heritage studies grapple with the international fair zones that display the massive machines and host the networks that make mega-mines and concrete jungles possible?

The rhetoric on the floor of the World Heritage Committee meetings, in the corridors of US embassies abroad, the conference rooms of the US State Department, and the various offices of ICOMOS and ICCROM would suggest that we have entered a period of wise-use, sustainable management. For the first time, the UN Sustainable Development Goals (UN Agenda 2030) include a specific call-out to cultural heritage. Target 11.4 reads: "Strengthen efforts to protect and safeguard the world's cultural and natural heritage." This would seem a good and reasonable goal given the last decade. The tragedies that have befallen archaeological and historic sites as well as museums at the hands of insurgent groups in the Middle East and North Africa region have raised global awareness such that one could reasonably expect commitments to and funding for collaborative research for the research, assessment, protection, and preservation of the human career. But, this hasn't happened.

What is more, the debates of heritage preservation in Turkey and the Middle East rarely encompass a deeper understanding of territorial modifications as they transcend ideological and territorial boundaries. The League of Nations and the mandates receive scrutiny for the arbitrary nature of lines drawn in the sand,[14] but the remaking of river basins along the Tigris in Iraq, the Jordan Valley, and the Litani in Lebanon, notably the drastic influence of UNESCO and the Arab-TVA,[15] has not been a topic of research in the context of cultural heritage per se. Even in areas where this confluence of actors has been recognized, such as the Nile,[16] and now also the Gediz Valley (this volume), the full scope of how heritage participates in water diplomacy schemes and development trends remains in its infancy.

In the case of western Turkey, it is not easy to break established patterns of agricultural practice on heritage landscapes. More than twenty-five years have passed since the initial Turkish implementation of heritage regulations in Bin Tepe and the Lake Marmara Basin. The inability to enforce current regulations and the unwillingness to engage in arbitration to reach suitable solutions demonstrates the vicious cycle of complacency and silent conflict. Over the last fifteen years of my research, members of the Izmir Protection Council and the Manisa Museum have each described this situation as a "Cold War" of colossal proportions among the sectors of extraction, water, agriculture, and archaeology.

Bureaucracies at all scales (i.e., from the offices of the AIA to various foreign ministries) are drowning in their own tortured web of mission statements, regulations, and policies. The leverage of private corporations to cut through the red tape of seemingly strong heritage regulations further emboldens their lust for power and weakens the mechanisms of heritage statecraft.[17] What is more, leveraging PPPs for infrastructure projects further distances the archaeology debate in countries with national ownership laws and ministries driven by the fear of plunder and the realities of duplicity. *Partage* remains a Pandora's box for consideration of serious foreign support. Arguably, Egypt would have never rallied the international attention it did without first promising temples (see Chapter 4, this volume). Salvage around the Tabqa dam in Syria in the 1960s, too, was also driven largely by concessions that granted rights of *partage*. Today countries sell marble and gold extracted from their landscapes, but cultural heritage is not for sale, nor for export.

From the perspective of many US museums and private philanthropists sponsorship of excavation policies of *partage* should be reinstated. Without these parameters, attracting funders is challenging. Currently the US Department of State offers no major support for research-driven archaeological fieldwork in Turkey, certainly not at the levels of the late 1950s and early 1960s. In fact, archaeological research is not considered a formal diplomatic tool. Official US policy concedes all such responsibility to the ivory tower. This issue became clear in the spring of 2014, when the Turkish Ministry of Culture and Tourism called for commitments from all foreign embassies that had active archaeological missions overseen by their citizens holding permits for excavation in Turkey. The US Embassy offered this statement to the American Research Institute in Turkey (ARIT):

> All American excavation teams are private, and have no administrative or financial relationship with the US federal government. In the American system, each American excavation should provide financial guarantees from its home institution or sponsors. The Embassy is normally given information about the activities of American archaeologists in Turkey, but has no responsibility for their finances or administrative operations.[18]

In addition, regardless of home institution and country of residence, citizenship dictates permit status and the benefits offered by the Ministry of Culture and Tourism.[19]

In spite of the prestige that academics, heritage practitioners, and diplomats often attach to heritage, especially if it carries the UNESCO brand, academic research for field archaeology remains impoverished. The US Ambassadors Fund for Cultural Preservation (AFCP) is another case in point. The US State Department's Cultural Heritage Center launched the fund in 2001 (before the September 11, 2001, terror attacks in the United States). It purports to support "the preservation of cultural sites, cultural objects, and forms of traditional cultural expression in more than 100 developing countries around the world."[20] On January 31, 2018, Acting Assistant Secretary for Education and Cultural Affairs Jennifer Zimdahl Galt made public remarks regarding crisis and heritage in Yemen. The backdrop for her call for heritage to contribute to stable governance, peace, and security was the Temple of Dendur at the Metropolitan Museum of Art in New York, "the gift" from Egypt for US patronage of Abu Simbel (see Chapter 4, this volume). In speaking about the dangerous networks associated with conflict antiquities, Galt hailed the importance of public events. She stressed that "[m]useums allow us to contemplate the role of history and culture in the development of modern society. Here in this magnificent space, we are inspired to consider our shared future and the heights of artistic and cultural endeavor that mankind can achieve." History tells us a different story, however. For some, the ongoing performance of the MET's Dendur platform represents not the hope of heritage, but rather a reminder of a once serious, sustained long-term support for not only monuments and objects, but also research and heritage landscapes.

In their call for the AFCP, the US Embassy in Turkey has considered proposals for the "preservation of an archaeological site," but they will not consider "[a]rchaeological excavations or exploratory surveys for research purposes" nor the "construction of new buildings, building additions, or permanent coverings (over archaeological sites, for example)." In addition, the AFCP may not support the "[r]elocation of cultural sites from one physical location to another."[21] In this context, US-supported projects such as the trainee sections of the Sardis excavation house, the Abu Simbel monuments, and the Temple of Dendur would not qualify. And, yet, they remain among the most steadfast components of active, in-country research, tourism, and backdrops for diplomatic performances.

Looking back to the early 1900s, a pattern emerges. Connections between philanthropic (private) and governmental sponsorship for archaeology were once much closer. The majority of the Sardis Society was Ivy League (notably Princeton men). The Wilsonian agenda was a significant part of the 1919 Paris Peace Conference, and Wilson's Fourteen Points offered inspiration for many educated elites who favored Hellenism (see Chapter 2, this volume). Visions that shaped the League of Nations and agendas undertaken by the ICIC during the interwar period, including the training of architects and urban planners in the renovation of ruined cities, but also conceptual ideas of how best to structure social and political life, moved away from a strict Classical perspective. Turkey differed from her southern and eastern neighbors in that the League of Nations issued no Mandate (the US having walked away). In retrospect, the country was able to balance a variety of different perspectives more easily. In the post-1946 world order the United Nations implemented broad sweeping programs tied to ideological agendas of neoliberalism. Over the years these types of projects increased in scope and number and an important sponsor (both directly and indirectly) was the US government.

In creating this type of structure, the US government endorsed a capitalist spirit. Archaeology in counties such as Turkey and Egypt offered a quick and relatively easy way to spend excess currency. The "democratization" that capitalist social mobility engendered opened up the field of educational inquiry that moved beyond a Classical foundation, and in doing so, the former close-knit, old-boy networks transformed. Today the predicament of heritage in limbo, from the facades of individual monuments to broad industrial landscapes, is how to re-envision its place in society and its long-term contribution. "The past for the future" has emboldened a consultancy culture that favors one-shot diplomatic deliverables at the expense of research. Without a serious, long-range perspective that is not only critical, but also reflexive, we face the risk of continuing to privilege heritage schemes devoid of holistic research. Understanding archaeology and heritage requires understanding the scale of the project and its agents, who balance local perspectives within the arc of global agendas.

NOTES

Introduction

1. Keeley, Edmund and Philip Sherrard, eds., *George Seferis Collected Poems.* Revised Edition (Princeton: Princeton University Press, 1995), 5.

2. The title of the song *Bakışları Hovarda, Buluşalım Fuarda* is difficult to translate into English. Something close to "Admire the big spenders, Let's meet at the Fair" captures the spirit. For the translation of the lyrics, I have relied on Sibel Bozdogan's *Modernism and Nation Building: Turkish Architectural Culture in the Early Republic* (Seattle: University of Washington Press, 2001). On page 141, she provides the Turkish text: Izmir'e bak, altından zincire bak / Akdeniz'in incise, üründe birincisi / Alıcı verici alsın, cihan hayrette kalsın / Türk işçisi baş alsın, cihan hayrette kalsın / Fuar dolup boşalsın, buluşalım Faur'da. From E. R. Üngör, *Türk Musikisi Güfteler Antolojisi*, (Istanbul: Eren Yayınları, 1980), 1280.

3. Interview with Wilbur P. Chase, by Charles Stuart Kennedy, July 24, 1990. From the Foreign Affairs Oral History Collection of the Association for Diplomatic Studies and Training.

4. See Meltem Eranıl Demirli, Zeynep Tuna Ultav, and Neslihan Demirtaş-Milz, "A Socio-Spatial Analysis of Urban Transformation at a Neighborhood Scale: The Case of the Relocation of Kadifekale Inhabitants to TOKİ Uzundere in Izmir," *Cities* 48 (2015): 140–159.

5. In reference to the name of the urban center, both Izmir and Smyrna were used throughout the nineteenth and early twentieth centuries. However, Izmir was favored among Muslim communities as well as used in Ottoman documents; foreigners (e.g., Europeans as well as Ottoman Greeks and Levantines) preferred Smyrna. Following Kolluoğlu Kırlı, I have chosen to use Smyrna for the Ottoman period, yet Izmir beginning with the Republic of Turkey. See Biray Kolluoğlu-Kırlı, "Forgetting the Smyrna Fire," *History Workshop Journal* 60, no. 1 (2005): 25–44.

6. For discussion of cultural life in Smyrna, see Edhem Eldem, Daniel Goffman, and Bruce Masters, *The Ottoman City between East and West: Aleppo, Izmir, and Istanbul* (New York: Cambridge University Press, 1999); and Daniel Goffman, *Izmir and the Levantine World: 1550–1650*, Vol. 5 (Seattle: University of Washington Press, 1990). For

discussion of Seferis' poetry, see Vayos Liapis, "'The Painful Memory of Woe': Greek Tragedy and the Greek Civil War in the Work of George Seferis," *Classical Receptions Journal* 6, no. 1 (2014): 74–103.

7. Key sources that focus on the 1922 fire and specifically the impact on the Greek and Armenian populations in Smyrna and Athens are as follows: Reşat Kasaba, "Izmir 1922: A Port City Unravels," in *Modernity & Culture: From the Mediterranean to the Indian Ocean*, ed. Leila Tarazi Fawaz and Christopher Alan Bayly (New York: Columbia University Press, 2002), 204–229; Kolluoğlu, "Forgetting the Smyrna Fire," 2005; and Leyla Neyzi, "Remembering Smyrna/Izmir: Shared History, Shared Trauma," *History & Memory* 20, no. 2 (2008): 106–127. George Horton's 1926 *Blight of Asia* did much to set a tone against Ottoman policies and Muslim communities; for a critical analysis see Corinna Tsakiridou, "Ideology, Prejudice, and Ethnic Conflict in George Horton's 'The Blight of Asia': A Radical Narrative Reconsidered," *Turkish Studies Association Bulletin* 25, no. 1 (2001): 21–38.

8. Kolluoğlu, "Forgetting the Smyrna Fire," 25; Philip Mansel, "Cities of the Levant—The Past for the Future?" *Asian Affairs* 45, no. 2 (2014): 220–242; and Ellinor Morack, "Fear and Loathing in 'Gavur' Izmir: Emotions in Early Republication Memories of the Greek Occupation (1919–22)," *International Journal of Middle East Studies* 49, no. 1 (2017): 71–89.

9. Sarah Abrevaya Stein, *Extraterritorial Dreams: European Citizenship, Sephardi Jews, and the Ottoman Twentieth Century* (Chicago: University of Chicago Press, 2016). For discussion of Jewish synagogues and funding, see Christina Luke, "Cultural Sovereignty in the Balkans and Turkey: The Politics of Preservation and Rehabilitation," *Journal of Social Archaeology* 13, no. 3 (2013): 350–370.

10. In 2015 a new international fair zone opened some distance from the city center. The new location is much closer to the Izmir Adnan Menderes International Airport and its Freeport zone. Between 2001 and 2017 I attended fairs in both locations.

11. Yaşar Akosy, *Love for 70 Years: Izmir Fair* (Izmir: Izmir Büyükşehir Belediyesi, 2001). The opening day of annual fairs throughout the 1930s was on September 9.

12. Faik Gur, "Sculpting the Nation in Early Republican Turkey," *Historical Research* 86, no. 232 (2013): 342–372.

13. Meskell, Lynn *A Future in Ruins: UNESCO, World Heritage, and the Dream of Peace*. London: Oxford University Press, 2018. Neil Brodie, "Syria and Its Regional Neighbors: A Case of Cultural Property Protection Policy Failure?" *International Journal of Cultural Property* 22, no. 2–3 (2015): 317–335; Susan Pollock, "Archaeology and Contemporary Warfare," *Annual Review of Anthropology* 45 (2016): 215–231; Claire Smith, Heather Burke, Cherrie de Leiuen, and Gary Jackson, "The Islamic State's Symbolic War: Da'esh's Socially Mediated Terrorism as a Threat to Cultural Heritage," *Journal of Social Archaeology* 16, no. 2 (2016): 164–188.

14. For discussion of the political discourse throughout the eastern Mediterranean and the "Eastern Question," see Michelle Tusan, *Smyrna's Ashes: Humanitarianism, Genocide, and the Birth of the Middle East* (Los Angeles: University of California Press, 2012).

15. The literature is expanding rapidly on this subject; see Feyzi Baban, Suzan Ilcan, and Kim Rygiel, "Syrian Refugees in Turkey: Pathways to Precarity, Differential Inclusion, and Negotiated Citizenship Rights," *Journal of Ethnic and Migration Studies* 43, no. 1 (2017): 41–57; and Ahmet Içduygu and Doğuş Şimşek, "Syrian Refugees in Turkey: Towards Integration Policies," *Turkish Policy Quarterly* 15, no. 3 (2016): 59–69.

16. For general discussion see, Ammar Kourany and Martha Myers, "Assistance Funding to Syria: For Development or Strife?" *Middle East Policy* 24, no. 2 (2017): 122–136. For specifics on antiquities, see Neil Brodie and Isber Sabrine, "The Illegal Excavation and Trade of Syrian Cultural Objects: A View from the Ground," *Journal of Field Archaeology* 43, no. 1 (2018): 1–11.

17. Philip Oltermann and Constanze Letsch, "Turkey Recalls Ambassador after German MPs' Armenian Genocide Vote," *The Guardian*, June 2, 2016.

18. Christina Luke, "The 40th World Heritage Session in Istanbul, Turkey: A Reflection on the Legacies of Heritage Policy and Missed Mega-Heritage," *Journal of Field Archaeology* 41, no. 6 (2016): 641–644. See also, Meskell, *A Future in Ruins*.

19. Heghnar Zeitlian Watenpaugh, "Preserving the Medieval City of Ani: Cultural Heritage between Contest and Reconciliation." *Journal of the Society of Architectural Historians* 73, no. 4 (2014): 528–555. Watenpaugh discusses Anadolu Kültur. For discussion of Kavala's arrest, see Sedat Ergin, "Why Has Osman Kavala Been Arrested?" *Hürriyet Daily News*, November 4, 2017, http://www.hurriyetdailynews.com/opinion/sedat-ergin/why-has-osman-kavala-been-arrested-121883.

20. The scholarship on this topic is enormous. I have found the following lectures to be particularly enlightening: George Mavrogordatos and Ayhan T. Aktar, "The 1923 Exchange of Populations: An Ongoing Debate," *American School of Classical Studies at Athens*, January 15, 2013 (accessed January 20, 2018), http://www.ascsa.edu.gr/index.php/News/newsDetails/george-mavrogordatos-and-ayhan-t.-aktar-the-1923-exchange-of-populations.-a.

21. Antonis Liakos, "Historical Time and National Space in Modern Greece," in *Regions in Central and Eastern Europe: Past and Present*, eds. Tadayuki Hayashi and Hiroshi Fukuda (Hokkaido: Hokkaido University, 2007), 205–227.

22. Seth P. Tillman, *Anglo-American Relations at the Paris Peace Conference of 1919* (Princeton, NJ: Princeton University Press, 2015), 365–368.

23. Corinna Tsakiridou, "Ideology, Prejudice and Ethnic Conflict in George Horton's 'The Blight of Asia': A Radical Narrative Reconsidered," *Turkish Studies Association Bulletin* 25, no. 1 (2001): 21–38.

24. Howard C. Butler, "Protection for the Historic Monuments and Objects of Art in Nearer Asia," *The Art Bulletin* 2, no. 1 (1919): 46–48.

25. Here I have drawn on key scholarship that explores legacies of development, such as Malin Hasselskog, "Rwandan Developmental 'Social Engineering': What Does It Imply and How Is It Displayed?" *Progress in Development Studies* 15, no. 2 (2015): 154–169; Malin Hasselskog, Peter J. Mugume, Eric Ndushabandi, and Isabell Schierenbeck, "National Ownership and Donor Involvement: An Aid Paradox Illustrated by the Case of Rwanda," *Third World Quarterly* 38, no. 8 (2017): 1–15; and Festus C.R.A Ikeotuonye, "Lateral Shades of Social Engineering: A Critical Exploration of 'Interest Representation,' 'State' and 'Development,'" *Community Development Journal* 37, no. 1 (2002): 69–79. Heritage statecraft, too, is an emerging component in this field; see Gertjan Plets, "Ethno-Nationalism, Asymmetric Federalism and Soviet Perceptions of the Past: (World) Heritage Activism in the Russian Federation," *Journal of Social Archaeology* 15, no. 1 (2015): 67–93; Gertjan Plets, "Heritage Statecraft: When Archaeological Heritage Meets Neoliberalism in Gazprom's Resource Colonies, Russia," *Journal of Field Archaeology* 41, no. 3

(2016): 368–383; and Gertjan Plets, "Heritage Bureaucracies and the Modern Nation State: Towards an Ethnography of Archaeological Systems of Government," *Archaeological Dialogues* 23, no. 2 (2016): 193–213.

26. The work of Lynn Meskell has exposed the politics of UNESCO. A few key publications of hers include the following, Lynn Meskell, "Gridlock: UNESCO, Global Conflict and Failed Ambitions," *World Archaeology* 47, no. 2 (2015): 225–238; Lynn Meskell, "Transacting UNESCO World Heritage: Gifts and Exchanges on a Global Stage," *Social Anthropology* 23, no. 1 (2015): 3–21. Meskell, *A Future in Ruins.*

27. Hobsbawm, Eric, and Terence Ranger, eds., *The Invention of Tradition* (New York: Cambridge University Press, [1983] 2012).

28. Çiğdem Atakuman, "Value of Heritage in Turkey: History and Politics of Turkey's World Heritage Nominations," *Journal of Mediterranean Archaeology* 23, no. 1 (2010): 107–131; Atakuman, Çiğdem, "Cradle or Crucible: Anatolia and Archaeology in the Early Years of the Turkish Republic (1923–1938)," *Journal of Social Archaeology* 8, no. 2 (2008): 214–235; Tuğba Tanyeri-Erdemir, "Archaeology as a Source of National Pride in the Early Years of the Turkish Republic," *Journal of Field Archaeology* 31, no. 4 (2006): 381–393.

29. Research on world heritage in Turkey is growing. Representative recent work includes Sara Bonini Baraldi, Daniel Shoup, and Luca Zan, "Understanding Cultural Heritage in Turkey: Institutional Context and Organisational Issues," *International Journal of Heritage Studies* 19, no. 7 (2013): 728–748; Gülsel Çiftci, Gülay Çakır, and Ali Çakır, "Political Crisis' and Their Implications on Heritage Sites: The Case of Turkey," *Journal of Tourism Theory and Research* 2, no. 2 (2016): 126–138; Helen Human, "Ethical or Empty Gestures?: World Heritage Nominations in Turkey," in *Heritage in Action*, eds. Helaine Silverman, Emma Waterton, and Steve Watson (Cham: Springer International Publishing, 2017), 75–89; and Helen Human, "Democratising World Heritage: The Policies and Practices of Community Involvement in Turkey," *Journal of Social Archaeology* 15, no. 2 (2015): 160–183.

30. Catherine Dolan and Dinah Rajak, eds., *Anthropology of Corporate Social Responsibility*, Vol. 18 (New York: Berghahn Books, 2016).

31. Here I've drawn on the work of the following scholars: John Clarke, Dave Bainton, Noémi Lendvai, and Paul Stubbs, *Making Policy Move: Towards a Politics of Translation and Assemblage* (Bristol: Policy Press, 2015); Noemi Lendvai and Paul Stubbs, "Assemblages, Translation, and Intermediaries in South East Europe: Rethinking Transnationalism and Social Policy," *European Societies* 11, no. 5 (2009): 673–695; and Natasha Wunsch, "Beyond Instrumentalisation: NGO Monitoring Coalitions in Croatia, Montenegro, and Serbia," *East European Politics* 31, no. 4 (2015): 452–467.

32. Jo-Anne Pemberton, "The Changing Shape of Intellectual Cooperation: From the League of Nations to UNESCO," *Australian Journal of Politics & History* 58, no. 1 (2012): 34–50.

Chapter 1

1. This quote is taken from the vision and mission tab of the Turkish Goldminers' Association website, accessed January 24, 2018, http://turkishgoldminersassociation.org/about/vizyon/.

2. Just past security in JFK's Terminal 1 stands a commemorative billboard: "In recognition of the generosity of the Republic of Turkey and the Aegean Minerals Exporters' Association for Izmir-Turkey for the 6,700 square meters of the original 'Turkish Burdur Beige Stone' floor tile which was donated."

3. The Turkish text on the billboard is as follows: "Türkiye'nin altın geleceği için çalışıyoruz. Tüprag, Metal Madencilik A.Ş. 1986 yılından bu yana gelişmiş teknoloji ve uzman ekibiyle ülke ekonomisine kazandırıyor." And, for the goldbar, "doğaya saygılı madencilik."

4. I was not paid for this work, nor, to my knowledge, were my colleagues.

5. World Heritage and Extractive Industries, accessed April 25, 2017, http://whc.unesco.org/en/extractive-industries/.

6. "Gölmarmara Hacıveliler Köyü Mermer Ocağı İstemiyor!" direnişteyiz (March 3, 2015); "Manisa'da köylüler mermer ocağına karşı nöbette [Villagers in Manisa on Guard Against Marble Quarry]," *Sol* (March 3, 2015).

7. See Nergis Canefe, "The Politics of Public Protests against Extractivism in Turkey," *Peace Review* 28, no. 1 (2016): 10–19.

8. "Haziran Salihli Madene Karşı Hacıveliler Köylüleriyle," *Muhalefet*, March 12, 2015.

9. Tim Arango, Kareem Fahim, and Sebnew Arsu, "Public's Outrage over Mine Disaster Casts Harsh Light on Turkey's Premier," *New York Times*, May 16, 2014. Kareem Fahim and Sebnem Arsu, "As Turkish Mine Yields the Last Body, Recriminations Remain," *New York Times*, May 17, 2014; Suzy Hansen, "The Mine Disaster That Shook Turkey," *New York Times Magazine*, November 26, 2014; Ceylan Yeginsu, "Anger and Grief Simmer in Turkey a Year after Soma Mine Disaster," *New York Times*, June 2, 2015.

10. See Fatih Demiroz and Naim Kapucu, "The Soma Mine Accident Case, Turkey," in *Routledge Handbook of Global Public Policy and Administration,* ed. Thomas R. Klassen, Denita Cepiku, and T. J. Lah (New York: Routledge, 2016), 207–216.

11. Christopher H. Roosevelt and Christina Luke, "The Story of a Forgotten Kingdom? Survey Archaeology and the Historical Geography of Central Western Anatolia in the Second Millennium BC," *European Journal of Archaeology* 20, no. 1 (2017): 120–147.

12. For details regarding the archaeological zoning and research at Kaymakçı, see İl Kültür ve Turizm Müdürlüğünün 18.03.2015 tarih ve 1156 sayılı yazısı; İl Kültür ve Turizm Müdürlüğünün 08.04.2015 tarih ve 1453 sayılı yazısı; Manisa M üze Müdürlüğünün 13.04.2015 tarih ve 81 O sayılı yazısı; İl Kültür ve Turizm Müdürlüğünün 28.05.2015 tarih ve 2159 sayılı yazısı; Kültür Varlıklan ve Müzeler Genel Müdürlüğünün 23.06.2015 tarih ve 121906 sayılı yazısı. Final verdict is Izmir 2 Numarlı Kültür Varlıklarını Koruma Bögle Kurulu Karar No. 29.04.28.6603. Examples of other recent updates to zoning reflective of submissions by the Central Lydia Archaeological Survey are the following: for tumuli, Izmir 2 Numarlı Kültür Varlıklarını Koruma Bögle Kurulu Karar No. 14.02.2014.3415, and Izmir 2 Numarlı Kültür Varlıklarını Koruma Bögle Kurulu Karar No. 11.04.2014.3745; Karakoyu, Izmir 2 Numarlı Kültür Varlıklarını Koruma Bögle Kurulu Karar No. 27.06.214.4052; Gedevre Tepsi, Izmir 2 Numarlı Kültür Varlıklarını Koruma Bögle Kurulu Karar No. 27.06.2014.4052; Kılcanlar, Izmir 2 Numarlı Kültür Varlıklarını Koruma Bögle Kurulu Karar No. 24.07.2015.5494-95; Kılcanlar tumuli, Izmir 2 Numarlı Kültür Varlıklarını Koruma Bögle Kurulu Karar

No. 24.07.2015.5496-97; Asartepe, Izmir 2 Numarlı Kültür Varlıklarını Koruma Bögle Kurulu Karar No. 24.07.2015.5492-93.

13. Bradley M. Sekedat, "X-Ray Fluorescence and Stable Isotope Analysis of Marble in Central Lydia, Western Turkey," *Oxford Journal of Archaeology* 35, no. 4 (2016): 369–388.

14. Christina Luke and Christopher H. Roosevelt, "Cup-Marks and Citadels: Evidence for Libation in 2nd-Millennium BCE Western Anatolia," *Bulletin of the American Schools of Oriental Research* 378 (2017): 1–23. For the importance of ancestral waterscapes and Bin Tepe, see Christina Luke and Christopher H. Roosevelt, "Memory and Meaning in Bin Tepe, the Lydian Cemetery of the 'Thousand Mounds,'" in *Tumulus as Sema: Space, Politics, Culture and Religion in the First Millennium BC* 27, ed. Olivier Henry, Ute Kelp (Boston: De Gruyter, 2016): 407–428.

15. For scholarship in this field, yet not specific to Turkey, see Elizabeth Ferry, "Gold Prices as Material-Social Actors: The Case of the London Gold Fix," *The Extractive Industries and Society* 3, no. 1 (2016): 82–85.

16. "Mining Turkey," *Mining & Earth Sciences Magazine* 5, no. 9 (2015): 1–68.

17. For scholarship on gold and Sardis, see Andrew Ramage, Paul T. Craddock, and Michael R. Cowell, *King Croesus' Gold: Excavations at Sardis and the History of Gold Refining*, Archaeological Exploration of Sardis Monograph. 11 (London: British Museum Publications, 2000).

18. See Michael L. Dougherty, Sarah Anderson, and Manuel Perez-Rocha, "The Rise of the Corporate Investment Rights Regime and 'Extractive Exceptionalism': Evidence from El Salvador," in *Mining in Latin America: Critical Approaches to the New Extraction*, ed. K. Deonandan and M. L. Dougherty (London: Routledge, 2016), 226–249.

19. Matthew Libassi and Nancy Lee Peluso, "Undergrounds above Ground: Four Views of Mining, Development and Society," *Development and Change* 47, no. 5 (2016): 1180–1195; Maria Lourdes Melo Zurita, Paul George Munro, and Donna Houston, "Un-earthing the Subterranean Anthropocene," *Area* (August 10, 2017), doi:10.1111/area.12369.

20. Ricoeur, Paul. "Memory and Forgetting," *Questioning Ethics: Contemporary Debates in Philosophy* (1999): 5–11. See also David Wood, ed., *On Paul Ricoeur: Narrative and Interpretation* (New York: Routledge, 2002).

21. Elena Frangakis-Syrett, "British Economic Activities in Izmir in the Second Half of the Nineteenth and Early Twentieth Centuries," *New Perspectives on Turkey* 6 (1991): 191–227.

22. Andrew J. Crozier, "The Establishment of the Mandates System 1919–25: Some Problems Created by the Paris Peace Conference," *Journal of Contemporary History* 14, no. 3 (1979): 483–513; Alfred E. Eckes, Jr., *The United States and the Global Struggle for Minerals* (Austin: University of Texas Press, 1979); Vincent W. Beach, "Exploiting the Satellites," *Queen's Quarterly* 64, no. 1 (1957): 64; E. T. Williams, "The Conflict between Autocracy and Democracy," *American Journal of International Law* 32, no. 4 (1938): 663–679.

23. Elena Frangakis-Syrett, "The Economic Activities of the Greek Community of Izmir in the Second Half of the Nineteenth and Early Twentieth Centuries," *Ottoman Greeks in the Age of Nationalism: Politics, Economy, and Society in the Nineteenth Century* (1999): 17–44; Elena Frangakis-Syrett, "Capital Accumulation and Family

Business Networks in Late Ottoman Izmir," *International Journal of Turkish Studies* 21, no. 1/2 (2015): 55–72.

24. William Hale, "Ideology and Economic Development in Turkey 1930–1945," *British Journal of Middle Eastern Studies* 7, no. 2 (1980): 100–117; H. Çağatay Keskinok, "Urban Planning Experience of Turkey in the 1930s," *METU JFA* 27, no. 2 (2010): 173–188.

25. Başak Kuş, "Weak States, Unruly Capitalists, and the Rise of Étatism in Late Developers: The Case of Turkey," *British Journal of Middle Eastern Studies* 42, no. 3 (2015): 358–374; Abdullah Takim and Ensar Yilmaz, "Economic Policy during Atatürk's Era in Turkey (1923–1938)," *African Journal of Business Management* 4, no. 4 (2010): 549.

26. Berch Berberoglu, "State Capitalism and National Industrialization in Turkey," *Development and Change* 11, no. 1 (1980): 97–121; Ömer Celâl Sarc, "Economic Policy of the New Turkey," *Middle East Journal* 2, no. 4 (1948): 430–446. Hale, "Ideology and Economic Development": 100–117.

27. Harry Flam, "Turkey and the EU: Politics and Economics of Accession," *CESifo Economic Studies* 50, no. 1 (2004): 171–210; Nahide Konak, "Ecological Modernization and Eco-Marxist Perspectives: Globalization and Gold Mining Development in Turkey," *Capitalism Nature Socialism* 19, no. 4 (2008):107–130, 118. Sübidey Togan, "Turkey: Toward EU Accession," *World Economy* 27, no. 7 (2004): 1013–1045.

28. Canefe, "The Politics of Public Protests": 14. Konak, "Ecological Modernization and Eco-Marxist Perspectives": 117.

29. Duygu Avcı, "Mining Conflicts and Transformative Politics: A Comparison of Intag (Ecuador) and Mount Ida (Turkey) Environmental Struggles," *Geoforum* 84 August (2017): 316–325. For larger discussions of extraction, see Andrew Barry, *Material Politics: Disputes along the Pipeline* (West Sussex: John Wiley & Sons, 2013); Catherine Dolan and Dinah Rajak, eds., *Anthropology of Corporate Social Responsibility*, Vol. 18 (New York: Berghahn Books, 2016). Elisa Giuliani, "Human Rights and Corporate Social Responsibility in Developing Countries' Industrial Clusters," *Journal of Business Ethics* 133, no. 1 (2016): 39–54; Peter T. Robbins and Ben Crow, "Engineering and Development: Interrogating Concepts and Practices," *Journal of International Development* 19, no. 1 (2007): 75–82; Bernhard Truffer and Lars Coenen, "Environmental Innovation and Sustainability Transitions in Regional Studies," *Regional Studies* 46, no. 1 (2012): 1–21.

30. The World Bank, "Using Data to Tell the Artisanal and Small-Scale Mining Story," January 30, 2017, http://www.worldbank.org/en/news/feature/2017/01/30/using-data-to-tell-the-artisanal-and-small-scale-mining-story.

31. Uğur Emek, "Turkish Experience with Public Private Partnerships in Infrastructure: Opportunities and Challenges," *Utilities Policy* 37 (2015): 120–129. Uğur Emek and Muhittin Acar, "6. Public Procurement in Infrastructure: The Case of Turkey," *Government Favouritism in Europe* (2015): 84.

32. Lindsay Shade, "Sustainable Development or Sacrifice Zone? Politics below the Surface in Post-Neoliberal Ecuador," *The Extractive Industries and Society* 2, no. 4 (2015): 775–784.

33. Rosemary J. Coombe and Melissa F. Baird, "The Limits of Heritage: Corporate Interests and Cultural Rights on Resource Frontiers," in *A Companion to Heritage*

Studies, 1st edition, eds. William Logan, Mairead Nic Craith, and Ullrich Kockel (West Sussex: Wiley & Sons Publishing, 2016), 337–354; Rosemary J. Coombe and Lindsay M. Weiss, "Neoliberalism, Heritage Regimes, and Cultural Rights," West Sussex: *Global Heritage: A Reader* (Wiley-Blackwell, 2015), 43–69; Melissa F. Baird, "Heritage, Human Rights, and Social Justice," *Heritage & Society* 7, no. 2 (2014): 139–155.

34. Coombe and Baird, "The Limits of Heritage," 339.

35. UN Global Compact, accessed March 10, 2017, https://www.unglobalcompact.org/what-is-gc.

36. Barry, *Material Politics*.

37. See Dolan and Rajak, *Anthropology of Corporate Social Responsibility*.

38. Frangakis-Syrett, "British Economic Activities in Izmir"; Elena Frangakis-Syrett, "Banking in Izmir in the Early Twentieth Century," *Mediterranean Historical Review* 24, no. 2 (2009): 115–132; Feryal Tansuğ, "The Greek Community of Izmir/Smyrna in an Age of Transition: The Relationship between Ottoman Centre-Local Governance and the Izmir/Smyrna Greeks, 1840–1866," *British Journal of Middle Eastern Studies* 38, no. 01 (2011): 41–72; Ertan Gökmen, "1888 Tarihli Akhisar Zımpara Madeni İmtiyazı" *Sosyal Bilimler* 2, no. 1 (2004): 53–64.

39. Erdoğan Atmiş, Sezgin Özden, and Wietze Lise, "Public Participation in Forestry in Turkey," *Ecological Economics* 62, no. 2 (2007): 352–359.

40. Cantürk Gümüş, "Historical Development of Forestry Education in the Context of Forest Resources Management in Turkey," *Turkish Journal of Forestry* 17, no. 1 (2016): 93–98.

41. In 1924, Law No. 474 began the cadastral process for Republic; in 1925, Law No. 658 established a cadastral section in the General Directorate of Land Registry; in 1934 the Cadastre and Land Registry Law was implemented, yet regulations were not prepared until 1935 and 1937 No. 3116.

42. Tahsin Boztoprak, Osman Demir, and Yakup Emre Çoruhlu, "Comparison of Expropriation and Land Consolidation on the Regulation of Agricultural Land," *Sigma* 34, no. 1 (2016): 43–55.

43. Osman Demir, Bayram Uzun, and Mehmet Çete, "Turkish Cadastral System," *Survey Review* 40, no. 307 (2008): 54–66.

44. The European Court of Human Rights (ECHR) has pursued cases of unjust expropriation, such as those that did not result in fair compensation. See Murat Volkan Dülger, "The Balance between the Public Interest and the Protection of Property Rights: A Judgment by the European Court of Human Rights about Expropriation," February 5, 2015, accessed January 2017 from the Social Science Research Network, https://ssrn.com/abstract=2560894.

45. Cihan Erdönmez and Aynur Aydin Coşkun, "Sustainability in Turkish Forest Legislation and Administration," accessed April 2017 from the Food and Agriculture Organization of the United Nations (FAO), http://www.fao.org/docrep/ARTICLE/WFC/XII/0072-C1.HTM; Ozge Balkiz, *Assessment of the Socio-Economic Values of Goods and Services Provided by Mediterranean Forest Ecosystems: Düzlerçami Forest, Turkey.* (Valbonne: Plan Bleu, 2016).

46. For broad discussion, see Nahide Konak, "Ecological Modernization and Eco-Marxist Perspectives: Globalization and Gold Mining Development in Turkey," *Capitalism Nature Socialism* 19, no. 4 (2008): 107–130.

47. İlknur Özgen, Jean Öztürk, and Machteld Johanna Mellink, *The Lydian Treasure: Heritage Recovered*, published by Uğur Okman for the Republic of Turkey, Ministry of Culture, General Directorate of Monuments and Museums, 1996. See also Morag M. Kersel, Christina Luke, and Christopher H. Roosevelt, "Valuing the Past Perceptions of Archaeological Practice in Lydia and the Levant," *Journal of Social Archaeology* 8, no. 3 (2008): 298–319; Christopher H. Roosevelt and Christina Luke, "Looting Lydia," *Archaeology, Cultural Heritage, and the Antiquities Trade, Gainesville, University Press of Florida* (2006): 173–187. Christopher H. Roosevelt and Christina Luke, "Mysterious Shepherds and Hidden Treasures: the Culture of Looting in Lydia, Western Turkey," *Journal of Field Archaeology* (2006): 185–198.

48. Ramage, Craddock, and Cowell, *King Croesus' Gold.*

49. Hayriye Özen and Şükrü Özen, "Interactions in and between Strategic Action Fields: A Comparative Analysis of Two Environmental Conflicts in Gold-Mining Fields in Turkey," *Organization & Environment* 24, no. 4 (2011): 343–363; Hayriye Özen and Şükrü Özen, "Public Policies and Social Movements: The Influences of Protest Movements on Mining Policy in Turkey," *Review of Public Administration* 43, no. 2 (2010): 33–64. Şükrü Özen and Hayriye Özen, "Peasants against MNCs and the State: The Role of the Bergama Struggle in the Institutional Construction of the Gold-Mining Field in Turkey," *Organization* 16, no. 4 (2009): 547–573.

50. This pit is enormous. The viewing platform is at the top, and from this vantage point, the massive scale of the operation becomes clear.

51. The lime is extracted locally from a limestone quarry in Uşak. This form of secondary extraction further enhances local economic opportunities, which then can be measured in terms of social mobility, and evidence of the "multipler effect."

52. Burcu Unsal and Hasan Yazicigil, "Assessment of Open Pit Dewatering Requirements and Pit Lake Formation for the Kışladağ Gold Mine, Uşak, Turkey." *Mine Water and the Environment* 35, no. 2 (2016): 180–198.

53. Özen and Özen, "Interactions in and between Strategic Action Fields": 351.

54. Gayrimenkul Eski Eserler Anıtlar Yüksek Kurulunun (GEEAYK)/ Real Estate Ancient Works Monument High Council 10.03.1978 No. A-1017; Second Izmir Protection Council 13.02.1991 No. 1815.

55. Sartlılar da Bergamalılar'ın izinde, *Dünya Gazetesi*, 22 Eylül 2010.

56. No. 45.07/16, December 8, 2016, meeting no. 208. Decision 08.12.2016-7394. Second Izmir Protection Council. Recent (2013 to present) Protection Council decisions can be found at the following link: http://www.korumakurullari.gov.tr/TR,90402/tescil-kararlari.html.

57. These include correspondences from the General Directorate, No. 174900 dated October 3, 2016, that updates the border zones, through official cadastral registration. This decision replaces those on file from the GEEAYK's (Gayrimenkul Eski Eserler Anıtlar Yüksek Kurulunun) decision of March 10, 1978 (decisions A-1017), the Second Protection Council of Izmir decision February 3, 1991 (No. 1814), and the Second Protection Council of Izmir's decision November 7, 2014 (No. 4474).

58. Christina Luke and Elvan Cobb, "Dwelling in Hacıveliler: Social-Engineering Policies in the Context of Space, Place and Landscape in Rural, Western Turkey," *Anatolian Studies* 63 (2013): 155–173.

59. Fiona Starr, *Corporate Responsibility for Cultural Heritage: Conservation, Sustainable Development, and Corporate Reputation*, Vol. 4. (New York: Routledge, 2013).

60. See the following correspondences: İl Kültür ve Turizm Müdürlüğünün 16.09.2011 tarih ve 169 sayılı yazısı; İl Kültür ve Turizm Müdürlüğünün 29.03.2013 tarih ve 1074 sayılı yazısı; İl Kültür ve Turizm Müdürlüğünün 02.03.2015 KAP tarihli dikelç; İl Kültür ve Turizm Müdürlüğünün 06.03.2015 tarihli ve 694 sayılı yazınız ve eki; İl Kültür ve Turizm Müdürlüğünün 18.03.2015 tarih ve 1156 sayılı yazısı; İl Kültür ve Turizm Müdürlüğünün 08.04.2015 tarih ve 1453 sayılı yazısı; İl Kültür ve Manisa Müze Müdürlüğünün 13.04.2015 tarih ve 810 sayılı yazısı; İl Kültür ve Turizm Müdürlüğünün 28.05.2015 tarih ve 2159 sayılı yazısı; Kültür Varlıkları ve Müzeler Genel Müdürlüğünün 23.06.2015 tarih ve 121906 sayılı yazısı.

61. See Decisions 29.04.2016-6603, 01.12.2011-278, and 28.06.2013-2498.

62. Ananya Roy, *Poverty Capital: Microfinance and the Making of Development* (London: Taylor and Francis, 2010).

Chapter 2

1. For discussion of Smyrna's varied communities, see Edhem Eldem, "The Undesirables of Smyrna, 1926," *Mediterranean Historical Review* 24, no. 2 (2009): 223–227; Athanasios Gekas, "Class and Cosmopolitanism: The Historiographical Fortunes of Merchants in Eastern Mediterranean Ports," *Mediterranean Historical Review* 24, no. 2 (2009): 95–114; Daniel Goffman, "Jews in Early Modern Ottoman Commerce," *Jews, Turks and Ottomans* (2002): 15–34; Ioanna Pepelasis Minoglou, "Ethnic Minority Groups in International Banking: Greek Diaspora Bankers of Constantinople and Ottoman State Finances, c. 1840–81," *Financial History Review* 9, no. 02 (2002): 125–146; and Sibel Zandi-Sayek, *Ottoman Izmir: The Rise of a Cosmopolitan Port, 1840–1880*. (Minneapolis: University of Minnesota Press, 2012).

2. The scholarship of Elena Frangakis-Syrett covers this topic very well. See Elena Frangakis-Syrett, "The Making of an Ottoman Port: The Quay of Izmir in the Nineteenth Century," *Journal of Transport History* 22, no. 1 (2001): 23–46; Elena Frangakis-Syrett, "American Trading Practices in Izmir in the Late Nineteenth and Early Twentieth Centuries," *Histoire économique et sociale de l'Empire ottoman et de la Turquie (1326–1960): actes du sixième congrès international tenu à Aix-en-Provence du 1er au 4 juillet 1992* [Economic and Social History of the Ottoman Empire and Turkey (1326–1960): Proceedings of the Sixth International Congress Held in Aix Provence from July 1st to 4th] 8 (1995): 177; Elena Frangakis-Syrett, "Implementation of the 1838 Anglo-Turkish Convention on Izmir's Trade: European and Minority Merchants," *New Perspectives on Turkey* 7 (1992): 91–112; Elena Frangakis-Syrett, "British Economic Activities in Izmir in the Second Half of the Nineteenth and Early Twentieth Centuries," *New Perspectives on Turkey* 6, Fall (1991): 191–227; and Elena Frangakis-Syrett, "Trade between Ottoman Empire and Western Europe: The Case of Izmir in the Eighteenth Century," *New Perspectives on Turkey* 2 (1988): 1–18.

3. For the first and only detailed study of Butler, see Fikret Yegül, "From the Lofty Halls of Academia to the Dusty Hills of Anatolia: Howard Crosby Butler and the First Sardis Expedition through Peace and War, 1909–1926," in *Perceptions of the Past in the Turkish Republic: Classical and Byzantine Periods*, eds. Scott Redford and Nina Ergin, *Ancient Near Eastern Studies*, Supplement 31 (Leuven, Belgium: Peeters, 2010), 57–100.

For Butler's work at Sardis, see Howard Crosby Butler. *Sardis I. The Excavations, Part 1: 1910–1914*. (Leiden: E. J. Brill, 1922).

4. Jack L. Davis and Natalia Vogeikoff-Brogan, eds., "Philhellenism, Philanthropy, or Political Convenience? American Archaeology in Greece," Special issue of *Hesperia: The Journal of the American School of Classical Studies at Athens* 82, no. 1 (2013); Jack L. Davis, "A Foreign School of Archaeology and the Politics of Archaeological Practice: Anatolia, 1922," *Journal of Mediterranean Archaeology* 16, no. 2 (2003): 145–172. For discussion of the founding of ARIT in 1964, and plans for something earlier (ca. 1920), see Christina Luke and Morag Kersel, *US Cultural Diplomacy and Archaeology: Soft Power, Hard Heritage* (New York: Routledge, 2013), 25–37.

5. Dockrill, Michael L., "Historical Note: The Foreign Office and the 'Proposed Institute of International Affairs 1919'," *International Affairs (Royal Institute of International Affairs 1944–)* 56, no. 4 (1980): 665–672.

6. Toynbee, Arnold. *The Western Question in Greece and Turkey* (London: Constable and Company, 1922).

7. Susan Pedersen, "Back to the League of Nations," *American Historical Review* 112, no. 4 (2007): 1091–1117; Susan Pedersen, *The Guardians: The League of Nations and the Crisis of Empire* (New York: Oxford University Press, 2015).

8. Patricia Clavin, *Securing the World Economy: The Reinvention of the League of Nations, 1920–1946* (New York: Oxford University Press, 2013); Paul Gordon Lauren, "Human Rights in History: Diplomacy and Racial Equality at the Paris Peace Conference," *Diplomatic History* 2, no. 3 (1978): 257–278; Erez Manela, *The Wilsonian Moment: Self-Determination and the International Origins of Anticolonial Nationalism* (Oxford University Press on Demand, 2007).

9. Louis W. Pauly, "Good Governance and Bad Policy: The Perils of International Organizational Overextension;" *Review of International Political Economy* 6, no. 4 (1999): 401–424; Louis W. Pauly, "The League of Nations and the Foreshadowing of the International Monetary Fund," *Essays in International Finance* 201 (1996): 1–47. See Thomas Davies, *NGOs: A New History of Transnational Civil Society* (New York: Oxford University Press, 2014).

10. For a discussion of heritage and property with regard to Sèvres, even if not specific to Izmir, see Ana Filipa Vrdoljak, "Self-Determination and Cultural Rights," In *Cultural Human Rights*, eds. Francesco Francioni and Martin Scheinin (Leiden, Netherlands: Brill, 2008), 41–78.

11. Erik Goldstein, "British Peace Aims and the Eastern Question: The Political Intelligence Department and the Eastern Committee, 1918," *Middle Eastern Studies* 23, no. 4 (1987): 419–436.

12. See also Matthias Bjørnlund, "The 1914 Cleansing of Aegean Greeks as a Case of Violent Turkification," *Journal of Genocide Research* 10, no. 1 (2008): 41–58.

13. For discussion of Toynbee, see Erik Goldstein, "The Foreign Office and Political Intelligence 1918–1920," *Review of International Studies* 14, no. 04 (1988): 275–288; Goldstein, "British Peace": 431.

14. Yegül, "From the Lofty Halls," 65–69.

15. Andrew Patrick, *America's Forgotten Middle East Initiative: The King-Crane Commission of 1919*. (London: I. B. Tauris, 2015); James Barr, *A Line in the*

Sand: Britain, France and the Struggle that Shaped the Middle East (New York: Simon and Schuster, 2011).

16. Here is the full text of Article 66 of the Treaty of Sèvres: ". . . thence northwards to a point to be chosen on the southern boundary of the Sandjak of Smyrna, a line to be fixed on the ground; thence to a point to be chosen in the neighbourhood of Bos Dagh situated about 15 kilometres northeast of Odemish, the southern and eastern boundary of the Sandjak of Smyrna; thence northwards to a point to be chosen on the railway from Manisa to Alashehr about 6 kilometres west of Salihli, a line to be fixed on the ground; thence northwards to Geurenez Dagh, a line to be fixed on the ground passing east of Mermer Geul west of Kemer, crossing the Kum Chai approximately south of Akshalan, and then followıng the watershed west of Kavakalan; thence northwestwards to a point to be chosen on the boundary between the Cazas of Kirkagach and Ak Hissar about 18 kilometres east of Kirkagach and 20 kilometres north of Ak Hissar, a line to be fixed on the ground."

17. Antonis Liakos, "Historical Time and National Space in Modern Greece," in *Regions in Central and Eastern Europe: Past and Present*, ed. Tadayuki Hayashi and Hiroshi Fukuda (Hokkaido: Hokkaido University, 2007), 205–227. See also Michael M. Finefrock, "Atatürk, Lloyd George and the Megali Idea: Cause and Consequence of the Greek Plan to Seize Constantinople from the Allies, June–August 1922," *Journal of Modern History* 52, no. S1 (1980): D1047–D1066.

18. For discussion of Classics in education and later business, see Victoria Tietze Larson, "Classics and the Acquisition and Validation of Power in Britain's 'Imperial Century' (1815–1914)," *International Journal of the Classical Tradition* 6, no. 2 (1999): 185–225. For discussion of Greek self-promotion through photography, see Katerina Zacharia, "Postcards from Metaxas' Greece: The uses of Classical Antiquity in Tourism Photography," in *Re-imagining the Past: Antiquity and Modern Greek Culture*, ed. Dēmētrēs Tziovas (New York: Oxford University Press), 187–208.

19. Corinna Tsakiridou, "Ideology, Prejudice and Ethnic Conflict in George Horton's 'The Blight of Asia': A Radical Narrative Reconsidered," *Turkish Studies Association Bulletin*, 25, no. 1 (2001): 21–38. Horton's wheeling and dealing eventually caught up with him: in 1923 he was removed from duty with the State Department. Yet, this couldn't fix the problems he'd created for Sardis by assisting with the illicit export of fifty-eight crates of material to the Metropolitan Museum of Art (MET). See Fikret Yegül, "Yağmalanan Sardes ve Türk Arkeolojisinin Geçmişle ilk Hesaplaşması, 1922–1925 [Plundered Sardis and a First Accounting of Turkish Archaeology's Past, 1922–1925]," in *Orhan Bingol 67. Yas Armagani*, ed. G. Kokdemir (Ankara, Turkey: Ankara University, 2013), 711–726.

20. See page 366 in Seth P. Tillman, *Anglo-American Relations at the Paris Peace Conference of 1919* (Princeton, NJ: Princeton University Press, 2015).

21. For a review of business interests, see John A. DeNovo, *American Interests and Policies in the Middle East, 1900–1939* (Minneapolis: University of Minnesota Press, 1963). See also the works of Elena Frangakis-Syrett. For discussion of Horton's role in trade and assistance for US citizens in Smyrna, see Donald Frothingham, "New Trade Route Open to Turkey: Big Help to US," *Chicago Daily Tribune* (April 6, 1914), 13.

22. Zeynep Çelik, *Displaying the Orient: Architecture of Islam at Nineteenth-Century World's Fairs*, Vol. 12. (Los Angeles: University of California Press, 1992); Robert W.

Rydell, *All the World's a Fair: Visions of Empire at American International Expositions, 1876–1916* (Chicago: University of Chicago Press, 2013).

23. Tony Bennett, *The Birth of the Museum: History, Theory, Politics* (New York: Routledge, 2013); Ingrid A. Steffensen-Bruce, *Marble Palaces, Temples of Art: Art Museums, Architecture, and American Culture, 1890–1930* (Lewisburg: Bucknell University Press, 1998). For the role of archaeology specifically, see Stephen L. Dyson, *Ancient Marbles to American Shores: Classical Archaeology in the United States* (Philadelphia: University of Pennsylvania Press, 1998).

24. Jack L. Davis, "The American School of Classical Studies and the Politics of Volunteerism," *Hesperia: The Journal of the American School of Classical Studies at Athens* 82, no. 1 (2013): 15–48.

25. Hastings worked with partner John M. Carrère to design the New York Public Library and the House and Senate Buildings in Washington, DC, among other buildings. William Charles Allen, *History of the United States Capitol: A Chronicle of Design, Construction, and Politics* (Washington, DC: U.S. Government Printing Office, 2001).

26. See Joseph L. Grabill, "Missionary Influence on American Relations with the Near East, 1914–1923," *The Muslim World* 58, no. 2 (1968): 141–154; and Joseph L. Grabill, "Cleveland H. Dodge, Woodrow Wilson, and the Near East," *Journal of Presbyterian History (1962–1985)* 48, no. 4 (1970): 249–264.

27. For scholarship on opium, Smyrna, and US interests, see Lawrence Karson, *American Smuggling as White Collar Crime* (New York: Routledge, 2014); Jacques M. Downs, "American Merchants and the China Opium Trade, 1800–1840," *Business History Review* 42, no. 04 (1968): 418–442; Geoffrey C. Ward, *Before the Trumpet: Young Franklin Roosevelt, 1882–1905* (New York: Vintage, 2014); and Martin Booth, *Opium: A History* (New York: St. Martin's Griffin, 2013). For specific linkages to Turkey, see Michael E. Chapman, "Taking Business to the Tiger's Gate: Thomas Handasyd Perkins and the Boston-Smyrna-Canton Opium Trade of the Early Republic," *Journal of the Royal Asiatic Society Hong Kong Branch* 52 (2012): 7–28; and Willem Floor, "From Anatolia to Indonesia: Opium Trade and the Dutch Community of Izmir, 1820–1940," *Journal of the American Oriental Society* 120, no. 4 (2000): 690–692. For the early trade in Smyrna and the importance of the Perkins family, see Timothy Mason Roberts, "Commercial Philanthropy: ABCfm Missionaries and the American Opium Trade," *The Role of the American Board in the World: Bicentennial Reflections on the Organization's Missionary Work, 1810–2010* (2012): 27. And, for the networks of families, notably the merger of Perkins to Russell and Company, see Kendall Johnson, "Henry James and the China Trade," *MFS Modern Fiction Studies* 60, no. 4 (2014): 677–710.

28. Steven Sora, *Secret Societies of America's Elite: From the Knights Templar to Skull and Bones* (Rochester, VT: Inner Traditions/Bear & Co., 2003).

29. See Gordon M. Winder, *The American Reaper: Harvesting Networks and Technology, 1830–1910* (London: Routledge, 2016); and Gordon M. Winder, "A Trans-National Machine on the World Stage: Representing McCormick's Reaper through World's Fairs, 1851–1902," *Journal of Historical Geography* 33, no. 2 (2007): 352–376.

30. Frangakis-Syrett, "American Trading Practices": 182–183; "Oppose America in Turkey,: German Diplomats Try to Prevent Grant of Asia Minor Concessions," *New York Times*, June 22, 1910. For a wider analysis of oil in this region and its social impact see

Teresa Brawner Bevis, *Higher Education Exchange between America and the Middle East through the Twentieth Century*. (New York: Palgrave Macmillan, 2016).

31. For Robert Garrett's background and interests, see Teresa T. Basler and David Wright, "The Making of a Collection: Mesoamerican Manuscripts at Princeton University," *Libraries & the Cultural Record* 43, no. 1 (2008): 29–55; and J. Michael Padgett, "The Collections of Ancient Art: The Early Years," *Record of the Art Museum, Princeton University* 55, no. 1/2 (1996): 107–124.

32. Director's Correspondence, Metropolitan Museum of Art, Microfilm Roll 2468. The Museum of Fine Arts (MFA) Boston archives demonstrate the careful paperwork that museums pursued to take advantage of favorable regulations pertaining to tax exemption for donations. They argued that taxes presented a hardship for public educational institutions.

33. Osman Hamdi Bey oversaw permits (*firman*) issued for excavations. In 1908 Buckler and Garrett invested $10,000 to begin the project and $8,000 per year thereafter for a total of $34,000. The Archaeological Institute of America (AIA) pledged $116,000. The total raised was $150,000. Sardis Excavation, 1908. Folder 15.3. AIA. For background on Hamdi Bey, see Ousterhout, Robert, and Renata Holod, eds. *Osman Hamdi Bey ve Amerikalılar Arkeoloji, Diplomasi, Sanat: Osman Hamdi Bey & The Americans Archaeology, Diplomacy, Art.* Pera Müzesi Yayınları. Sergi Kataloğu / Exhibition Catalogue 51. İstanbul: Suna ve İnan Kıraç Vakfı, Pera Müzesi, 2011.

34. See Padgett, "The Collections of Ancient Art": 115.

35. Edward Robinson to Francis Kelsey, May 5, 1908. Sardis Excavation. File 15.3. AIA.

36. Everit Macy to H. C. Butler, December 21, 1911. Howard Crosby Butler Papers (HCB), binder 15. Princeton University (PU).

37. Everit Macy to H. C. Butler, March 1, 1912. HCB, binder 19, PU. Everit Macy to H. C. Butler, May 15, 1912. HCB, folder 1121, PU.

38. Everit Macy to H. C. Butler, May 15, 1912. HCB, binder 21. PU.

39. For discussion of Christian Arabs, see I. Sardis, "The Excavations Part I, 1910–1914" (Leiden, Netherlands: Brill, 1922), 37. For discussion of Cretans and a Sephardic Jewish commissioner, see W. H. Buckler to Lucy Buckler, May 14, 1910, and June 8, 1910. William Hepburn Buckler Papers (WHB). MS 654. Series 1. Box 4, folder 46. Yale University (YU).

40. First applied in Syria and then at Sardis, individual architectural fragments were drawn in exquisite ornamental detail, fine lines calling attention to form and decoration. This approach followed a long tradition, from Stuart and Revett's publications on Athens and environs in the mid-eighteenth century to those of the Society of the Dilettanti. The American School for Classical Studies in Athens and the American Academy in Rome, too, endorsed this style. See James Stuart, Nicholas Revett, and Frank Salmon, *The Antiquities of Athens: Measured and Delineated by James Stuart, FRS and FSA, and Nicholas Revett, Painters and Architects* (New York: Princeton Architectural Press, 2008 [1762]); and Fikret K. Yegül, *Gentlemen of Instinct and Breeding: Architecture at the American Academy in Rome, 1894–1940* (New York: Oxford University Press, 1991).

41. Beyond their obvious professional relationship from fieldwork at Sardis, Butler's decision to appoint Buckler as the AIA representative, which he argued to Breasted at the

Oriental Institute in Chicago, was that Buckler was living in London, had connections through his half-brother, US diplomat Henry White, and was "an enthusiastic archaeologist in close touch with the British and French archaeologists." See H. C. Butler to James H. Breasted, December 1918. HCB, binder 77, PU. He also wrote to Robinson at the Metropolitan Museum of Art. H. C. Butler to Edward Robinson. December 1918. HCB, binder 80, PU.

42. W. H. Buckler to G. G. Buckler, March 31, 1919. WHB. MS 654. Box 1, folder 20. YU. The Annex to proposed Article 421 of Sèvres was as follows:

1. *"Antiquity" means any construction or any product of human activity earlier than the year 1700.*
2. *The law for the protection of antiquities shall proceed by encouragement rather than by threat. Any person who, having discovered an antiquity without being furnished with the authorisation referred to in paragraph 5, reports the same to an official of the competent Turkish Department, shall be rewarded according to the value of the discovery.*
3. *No antiquity may be disposed of except to the competent Turkish Department, unless this Department renounces the acquisition of any such antiquity. No antiquity may leave the country without an export licence from the said Department.*
4. *Any person who maliciously or negligently destroys or damages an antiquity shall be liable to a penalty to be fixed.*
5. *No clearing of ground or digging with the [objective] of finding antiquities shall be permitted, under penalty of fine, except to persons authorised by the competent Turkish Department.*
6. *Equitable terms shall be fixed for expropriation, temporary or permanent, of lands which might be of historical or archæological interest.*
7. *Authorisation to excavate shall only be granted to persons who show sufficient guarantees of archaeological experience. The Turkish Government shall not, in granting these authorisations, act in such a way as to eliminate scholars of any nation without good grounds.*
8. *The proceeds of excavations may be divided between the excavator and the competent Turkish Department in a proportion fixed by that Department. If division seems impossible for scientific reasons, the excavator shall receive a fair indemnity in lieu of a part of the find.*

43. H. C. Butler to Edward Robinson, December 1918. HCB, binder 80, PU.

44. Amended by Public Law 994, Chap. 982, 84th Congress—2nd Session, 70 Stat. 1060 (1956), approved by President Dwight D. Eisenhower on August 6, 1956.

45. For more on the history of the discipline, see Stephen L. Dyson, *Ancient Marbles to American Shores: Classical Archaeology in the United States* (Philadelphia: University of Pennsylvania Press, 1998).

46. Charles E. Norton, "The Work of the Archaeological Institute of America: An Address," *American Journal of Archaeology* 4, no. 1 (1900): 1–16.

47. G. M. Whicher to H. C. Butler, January 2, 1919. HCB, binder 35, PU. Report on the Proposals made to the Peace Conference with a view to placing the antiquities within the Ottoman Empire under the protection of the League of Nations. The Unofficial Drafting Committee, written by H. C. Butler. AIA.

48. G. M. Whicher to H. C. Butler, February 7, 1919. Paris. AIA. G. M. Whicher to H. C. Butler, March 11, 1919. AIA. The University Museum of Philadelphia and the Smithsonian Institution declined to endorse the initiative. For the University Museum of Philadelphia, see Harrison to AIA, January 21, 1919. AIA. For the Smithsonian Institution, see Smithsonian to AIA, February 11, 1919. AIA.

49. The ACLS gained full legal title only in 1920.

50. Significance of the "Eastern Establishment" in foundations and pursuit of archaeological research can be found in Thomas C. Patterson, "The Last Sixty Years: Toward a Social History of Americanist Archeology in the United States," *American Anthropologist* 88, no. 1 (1986): 7–26. For discussion of the immense power of US corporations, see Gabriel Kolko, *The Triumph of Conservatism: A Reinterpretation of American History, 1900–1918* (New York: Free Press, 1963); and James Weinstein, *The Corporate Ideal in the Liberal State: 1900–1918* (Boston: Beacon Press, 1968).

51. See Inderjeet Parmar, *Foundations of the American Century: The Ford, Carnegie, and Rockefeller Foundations and the Rise of American Power* (New York: Columbia University Press, 2013).

52. Buckler to Butler, December 1, 1919. The Unofficial Drafting Committee. Report on the proposals made to the Peace Conference with a view to placing the antiquities within the Ottoman Empire under the protection of the League of Nations. HCB, PU. See also Suggestions for the Internationalization of the Historic Monuments of Nearer Asia, HCB, PU.

53. W. H. Buckler, December 1, 1919. Report on the Proposals made to the Peace Conference with a View to Placing the Antiquities within the Ottoman Empire under the Protection of the League of Nations. American Commission to Negotiate Peace. Hotel Crillon, Paris. AIA.

54. For this type of organization, see Philip Marshall Brown, "From Sevres to Lausanne," *American Journal of International Law* 18, no. 1 (1924): 113–116.

55. Howard Crosby Butler, "Protection for the Historic Monuments and Objects of Art in Nearer Asia," *The Art Bulletin* 2, no. 1 (1919): 46–48.

56. Ibid, 48.

57. Report on the Proposals made to the Peace Conference with a view to placing the antiquities within the Ottoman Empire under the protection of the League of Nations. The Unofficial Drafting Committee, written by H. C. Butler. Approval by Union Academique Internationale. AIA. For his personal account of the meeting, see W. H. Buckler to G. G. Buckler, October 18, 1919. WHB. MS 654. Series 1, Box 2, Folder 23. YU.

58. All distances from Smyrna are from Alsancak.

59. See Benjamin Anderson, "'An Alternative Discourse': Local Interpreters of Antiquities in the Ottoman Empire," *Journal of Field Archaeology* 40, no. 4 (2015): 450–460; James F. Goode, *Negotiating for the Past: Archaeology, Nationalism, and Diplomacy in the Middle East, 1919–1941* (Austin: University of Texas Press, 2007); Morag M. Kersel, Christina Luke, and Christopher H. Roosevelt, "Valuing the Past Perceptions of Archaeological Practice in Lydia and the Levant," *Journal of Social Archaeology* 8, no. 3 (2008): 298–319; and Wendy M. K. Shaw, *Possessors and Possessed: Museums, Archaeology, and the Visualization of History in Ottoman Empire* (Los Angeles: University of California Press, 2003).

60. Edhem Eldem, "From Blissful Indifference to Anguished Concern: Ottoman Perceptions of Antiquities, 1799–1869," in *Scramble for the Past: A Story of Archaeology in the Ottoman Empire, 1914*, eds. Zainab Bahrani, Zeynep Çelik, and Edhem Eldem (Istanbul: SALT, 2011), 281–330.

61. For discussion of Kolophon and ASCSA, see Davis, "A Foreign School." For the occupation period (and thereafter the Early Republic), see Melania Savino, "Connecting Sites and Images: Archeology as Controversial Knowledge in Modern Izmir," *History of Science* 55, no. 3 (2017): 364–382.

62. For discussion of secret agreements and Venizelos' attempt to maintain power, see Goldstein, "The Foreign Office." For Buckler's reaction to secret agreements, see W. H. Buckler to G. G. Buckler, August 8, 1919. WHB. MS 654. Box 2, folder 22, YU. For discussion of Butler's visions of export, see Yegül, "From the Lofty Halls."

63. Notes on the Meeting of the Commissioners and the Technical Experts of the American Commission to Negotiate Peace, Hotel Crillon, Paris, Tuesday July 16, 1919. WHB. MS 654. Series III. Box 6, folder 3, YU. Discussions continued in September; see Notes on the Meeting of the Commissioners and the Technical Experts of the American Commission to Negotiate Peace, Hotel Crillon, Paris, Tuesday September 18, 1919. WHB. MS 654. Series III. Box 6, folder 3, YU.

64. W. H. Buckler, Memorandum on the Policy of the United States Relative to the Treaty with Turkey, November 26, 1919. WHB. MS 654. Series III. Box 6, folder 6b, YU. See also W. H. Buckler to G. G. Buckler, April 26, 1919. WHB. MS 654. Box 1, folder 20, YU.

65. A. L. Macfie, "The Revision of the Treaty of Sèvres: The First Phase (August 1920–September 1922)," *Balkan Studies* 24 (1983): 57–88.

66. Venizelos to Clemenceau. Letter from the Greek Delegation on the situation at Smyrna, and the censorship exercised by the Delegates of the Principal Allied and Associated Powers in that city. Greek Delegation to the Peace Congress, 4 November 1919. WHB. MS 654. Box 2, folder 21, YU.

67. American Council of Learned Societies: Notice of first Meeting. AIA.

68. Yegül's "From the Lofty Halls."

69. W. H. Buckler to G. G. Buckler, May 23, 1919. WHB. MS 654. Box 2, folder 21, YU.

70. W. H. Buckler to G. G. Buckler, June 5, 1919. WHB. MS 654. Box 2, folder 21, YU.

71. W. H. Buckler to G. G. Buckler, March 31, 1919. WHB. MS 654. Box 6, folder 2, YU.

72. A report entitled "Greek Atrocities in the Vilayet of Smyrna (May–July 1919)" was published by The Permanent Bureau of the Turkish Congress at Lausanne.

73. See the Circular of the Ottoman League, dated May 31, 1919, Sub. No 20, which asks "to draw once more and more seriously than ever, the attention of Europe and of America to a state of things likely to engender grave perturbations in the East." See also, the translation of the letter No 1/24516, dated 26th June 1919, addressed by the Colonel Kemal, Commandant General of the Gendarmerie, to the French Colonel Foulon, Inspector General: "unbearable calamities under which she is overwhelmed to point out these crimes to the powerful personages who hold in their hands the destinies of the

world and to inform the civilised universe that the Turks and the Mussulmans are being slaughtered like sheep and that their existence, their property and their honour are being annihilated. Your compassionate intervention will perhaps contribute to put an end to the injustices to which is exposed an unhappy people which, like every other nation has a right to existence."

74. Ellinor Morack, "Fear and Loathing in 'Gavur' Izmir: Emotions in Early Republication Memories of the Greek Occupation (1919–22)," *International Journal of Middle East Studies* 49, no. 1 (2017): 71–89.

75. See Toynbee, "The Western Question." A recent analysis of Arnold Toynbee's assessment can be found here: Rebecca Gill, "'Now I have seen evil, and I cannot be silent about it': Arnold J. Toynbee and His Encounters with Atrocity, 1915–1923," in *Evil, Barbarism and Empire* (New York: Palgrave Macmillan 2011), 172–200.

76. William Ramsay to H. C. Butler, April 27, 1918. HCB, binder 41, PU.

77. Karo to US Consul in Smyrna, June 21, 1918. Papers concerning destruction of Butler's dig house, 1918–1922, provided by C. H. Greenewalt. Copies are also kept by Princeton University, Cornell University, and Harvard University. See also Yegül, "From the Lofty Halls," footnote 37.

78. Swedish Consul Van der Lee to US Charge d'Affaires Colonel Ahlgran, Constantinople, July 24, 1918. Papers concerning the destruction of Butler's dig house, 1918–1922, provided by C. H. Greenewalt.

79. Correspondence between US commissioner in Constantinople, January 20, 1919, and US Secretary of State, Washington DC. Papers concerning destruction of Butler's dig house, 1918–1922, provided by C. H. Greenewalt.

80. The Commander of the Gendarmarie of Salehli, Tefik, 1st Lieutenant. Papers concerning destruction of Butler's dig house, 1918–1922, provided by C. H. Greenewalt.

81. US Assistant Secretary of State Robert Lansing to H. C. Butler, September 6, 1919. HCB, binder 130. PU.

82. Toynbee to Hill, February 10, 1921. HCB, binder 2, PU.

83. This group subsequently joined the Greek forces, and once in Smyrna, fled.

84. For detailed illustration of the advancement of Greek troops and the Turkish fronts, see Teoman Ergül, *Kurtuluş Savaşı'nda Manisa* (Ankara: Kebikçe, 2007). Greek maps also illustrate the movement of troops: Η ΕΚΣΤΡΑΤΕΙΑ ΕΙΣ ΤΗΝ ΜΙΚΡΑΝ ΑΣΙΑΝ 1957 = Η ΕΚΣΤΡΑΤΕΙΑ ΕΙΣ ΤΗΝ ΜΙΚΡΑΝ ΑΣΙΑΝ 1919–1922. ΓΕΝΙΚΟ ΕΠΙΤΕΛΕΙΟ ΣΤΡΑΤΟΥ, Athens 1957.

85. A. E. Montgomery, "The Making of the Treaty of Sèvres of 10 August 1920," *The Historical Journal* 15, no. 4 (1972): 786.

86. Ibid.

87. Goode, *Negotiating for the Past.*

88. H. C. Butler to S. A. Papafrango, December 14, 1921. HCB, binder 41. PU.

89. S. A. Papafrango to H. C. Butler, March 2,1922. HCB, binder 47, PU.

90. Fikret Yegül, "Architect Throwing Stone over Column 7 of Temple of Artemis," in *The City of Sardis: Approaches in Graphic Recording*, ed. Crawford H. Greenewalt (Cambridge, MA: Harvard University Art Museum, 2003), 74–75.

91. Yegül, "From the Lofty Hills."

92. H. C. Butler to Edward Capps, May 30, 1922. Papers concerning destruction of Butler's dig house 1918–1922, provided by C. H. Greenewalt.

93. Cyrus McCormick to H. C. Butler, April 19, 1922. HCB, binder 54, PU.

94. Edward Robinson to H. C. Butler, April 10, 1922. HCB, binder 52, PU.

95. H. C. Butler to William C. Osborn, April 18, 1922. HCB, binder 53, PU.

96. H. C. Butler to Edward Robinson, April 22, 1922. HCB, binder 57, PU.

97. S. C. H. Smith (Manager of Standard Oil Compnay of New York in Smyrna) to H. C. Butler, June 9, 1922. HCB, binder 60, PU.

98. H. C. Butler statement before Maynard B. Barnes Vice Consul of the United States of America at Smyrna, July 8, 1922. HCB, binder 64, PU. W. H. Buckler (Bill) to H. C. Butler (Chelibi), July 22, 1922. Papers concerning destruction of Butler's dig house 1918–1922, provided by C. H. Greenewalt.

99. Khānum Hālidah Adib, *The Turkish Ordeal* (New York: Century, 1928).

100. Goode, *Negotiating for the Past*.

101. Buckler to Butler, December 21, 1918. Letter of Correspondence and Resolution Passed by Hellenic Council. AIA archives: b018f14.

Chapter 3

1. Interview with Ambassador Gilbert A. Robinson, August 15, 2002, Association for Diplomatic Studies and Training Foreign Affairs Oral History Project.

2. Cyrus Veeser, "A Forgotten Instrument of Global Capitalism? International Concessions, 1870–1930," *International History Review* 35, no. 5 (2013): 1136–1155.

3. Christopher J. Barrow, "River Basin Development Planning and Management: A Critical Review," *World Development* 26, no. 1 (1998): 171–186.

4. US engineers from General Electric visited Russia to discuss the five-year Russian plans; they also consulted on the construction of the Dneiper. For discussion of how this work influenced the design of the New Deal, see Rosalyn McKeown-Ice, "Rostov Region, Russia-Tennessee Valley, USA Partnership for Training in Environmental Education and Management," *International Research in Geographical and Environmental Education* 3, no. 2 (1994): 77–80. See also Anne Dickason Rassweiler, *The Generation of Power: The History of Dneprostroi* (New York: Oxford University Press, 1988).

5. Sibel Bozdoğan, "Architecture, Modernism and Nation-Building in Kemalist Turkey," *New Perspectives on Turkey* 10 (1994): 37–55. See also Sibel Bozdoğan and Reşat Kasaba, eds. *Rethinking Modernity and National Identity in Turkey*, Vol. 7 (Seattle: University of Washington Press, 1997).

6. For discussion of the New Deal in the US, see Sarah T. Phillips, *This Land, This Nation: Conservation, Rural America, and the New Deal* (New York: Cambridge University Press, 2007).

7. Philip Selznick, *TVA and the Grass Roots: A Study in the Sociology of Formal Organization*, Vol. 3 (Berkeley: University of California Press, 1949).

8. Tim Culvahouse (ed.), *The Tennessee Valley Authority: Design and Persuasion*, Vol. 8. (New York: Princeton Architectural Press, 2007).

9. David Lilienthal to Julian Huxley, December 15, 1942. OGM. Tennesse Valley Authority Archives (TVA). For discussion of TVA on the Danube (DVA), see Vincent Lagendijk, "Divided Development: Post-War Ideas on River Utilisation and Their Influence on the Development of the Danube," *International History Review* 37, no. 1 (2015): 80–98.

10. Huxley, Julian. *TVA, Adventure in Planning*. (Kingston on Thames: Knapp, Drewett & Sons Ltd., 1943).

11. Patricia Clavin, *Securing the World Economy: The Reinvention of the League of Nations, 1920–1946* (New York: Oxford University Press, 2013); Patricia Clavin and Jens-Wilhelm Wessel, "Transnationalism and the League of Nations: Understanding the Work of Its Economic and Financial Organisation," *Contemporary European History* 14, no. 04 (2005): 465–492; Louis W. Pauly, *Who Elected the Bankers? Surveillance and Control in the World Economy* (Ithaca, NY: Cornell University Press, 1997); Frederick J. Peterson, "The Facade of Humanitarian Intervention for Human Rights in a Community of Sovereign Nations," *Arizona Journal of International and Comparative Law* 15 (1998): 871–904.

12. Representative publications include Mardges Bacon, *Le Corbusier in America: Travels in the Land of the Timid* (Cambridge, MA: MIT Press, 2001); Zeynep Celik, "Le Corbusier, Orientalism, Colonialism," *Assemblage* 17 (1992): 59–77; Enis Kortan, *Turkish Architecture and Urbanism through the Eyes of Le Corbusier* (Ankara: Middle East Technical University, 1991); Eric Mumford, "CIAM Urbanism after the Athens Charter," *Planning Perspective* 7, no. 4 (1992): 391–417; and Eric Paul Mumford, *The CIAM Discourse on Urbanism, 1928–1960* (Cambridge, MA: MIT Press, 2002).

13. For broad-scale discussion of Le Corbusier's approach and development theory, see James C. Scott, *Seeing Like a State: How Certain Schemes to Improve the Human Condition Have Failed* (New Haven, CT: Yale University Press, 1998). For general discussion of modernism, see John R. Gold, *The Experience of Modernism: Modern Architects and the Future City, 1928–53* (New York: Taylor & Francis, 2013); and Stephen V. Ward, "Soviet Communism and the British Planning Movement: Rational Learning or Utopian Imagining?" *Planning Perspectives* 27, no. 4 (2012): 499–524.

14. Konstanze Sylva Domhardt, "The Garden City Idea in the CIAM Discourse on Urbanism: A Path to Comprehensive Planning," *Planning Perspectives* 27, no. 2 (2012): 173–197.

15. Mumford, *The CIAM Discourse on Urbanism*, 45.

16. David Arredondo-Garrido, "References in Le Corbusier's Reorganization of Rural Habitat," *Journal of Architecture and Urbanism* 40, no. 2 (2016): 99–109; Mary McLeod, "The Country Is the Other City of Tomorrow? Le Corbusier's Ferme Radieuse and Village Radieux," in *Food and the City: Histories of Culture and Cultivation*, ed. Dorothée Imbert (Washington, DC: Dumbarton Oaks Research Library and Collection, 2015), 99–122.

17. Mardges Bacon, "Le Corbusier and Postwar America," *Journal of the Society of Architectural Historians* 74, no. 1 (2015): 13–40.

18. David E. Lilienthal, *Democracy on the March* (New York: Harper & Brothers, 1944). Lilienthal's work with the US government set the stage for the formation of his own company, Development and Resources Corporation, launched in 1955, which went on to win a number of US government contracts. See also Christopher T. Fisher, "'Moral Purpose Is the Important Thing': David Lilienthal, Iran, and the Meaning of Development in the US, 1956–63," *International History Review* 33, no. 3 (2011): 431–451.

19. David Ekbladh, "'Mr. TVA': Grass-Roots Development, David Lilienthal, and the Rise and Fall of the Tennessee Valley Authority as a Symbol for US Overseas Development, 1933–1973," *Diplomatic History* 26, no. 3 (2002): 335–374; Stephen Macekura, "The Point Four Program and US International Development Policy," *Political Science Quarterly* 128, no. 1 (2013): 127–160.

20. Selznick, *TVA and the Grass Roots.*

21. David Ekbladh, "Meeting the Challenge from Totalitarianism: The Tennessee Valley Authority as a Global Model for Liberal Development, 1933–1945," *International History Review* 32, no. 1 (2010): 47–67. By 1950 over 2,000 international visitors had come through TVA as part of the US development diplomacy; see *Congressional Record: Proceedings and Debates of the Congress*, Vol. 93, Part 7 (Washington, DC: US Government Printing Office, 1947).

22. Macekura, "The Point Four Program": 138.

23. Leila M. Harris and Samer Alatout, "Negotiating Hydro-Scales, Forging States: Comparison of the Upper Tigris/Euphrates and Jordan River Basins," *Political Geography* 29, no. 3 (2010): 148–156; Timothy Mitchell, *Rule of Experts: Egypt, Techno-Politics, Modernity* (Los Angeles: University of California Press, 2002); Chris Sneddon and Coleen Fox, "The Cold War, the US Bureau of Reclamation, and the Technopolitics of River Basin Development, 1950–1970," *Political Geography* 30, no. 8 (2011): 450–460.

24. See Andrew Barry, *Material Politics: Disputes along the Pipeline* (West Sussex: John Wiley & Sons, 2013); James Ferguson, *The Anti-Politics Machine: 'Development,' Depoliticization and Bureaucratic Power in Lesotho* (New York: Cambridge University Press, 1990). See also Mitchell, "*Rule of Experts*."

25. See the interviews archived in the US by the Foreign Affairs Oral History Collection of the Association for Diplomatic Studies and Training: 2002 with Ambassador Gilbert A. Robinson; 1998 with Ambassador William E. Rau; and 1990 with Ambassador Wilbur P. Chase. See also Corinna Tsakiridou, "Ideology, Prejudice and Ethnic Conflict in George Horton's 'The Blight of Asia': A Radical Narrative Reconsidered," *Turkish Studies Association Bulletin* 25, no. 1 (2001): 21–38.

26. Lyrics by Behçet Kemal Çağlar; composition by Ahmed Adnan Saygun. I thank Elif Doğan and Özge Ertem for their assistance with this translation. I also consulted that of Dr. Baykan Günay as cited by Ayşe Duygu Kaçar, "Cultivating the Nation: Atatürk's Experimental Farm as an Agent of Social and Cultural Transformation" (PhD dissertation, School of Natural and Applied Sciences, Middle Eastern Technical University, Ankara, Turkey, 2010.

27. Michael M. Finefrock, "Laissez-Faire, the 1923 Izmir Economic Congress and Early Turkish Developmental Policy in Political Perspective," *Middle Eastern Studies* 17, no. 3 (1981): 375–392.

28. Duygu Kacar, "A Unique Spatial Practice for Transforming the Social and Cultural Patterns: Atatürk Forest Farm in Ankara (1)," *METU JFA* 1 (2011): 147.

29. Şener Aktürk, "The Fourth Style of Politics: Eurasianism as a Pro-Russian Rethinking of Turkey's Geopolitical Identity," *Turkish Studies* 16, no. 1 (2015): 54–79; Sener Aktürk, "Politics of History in Turkey: Revisionist Historiography's Challenge to the Official Version of the Turkish War of Liberation (1919–1922)," in *The Convolutions of Historical Politics*, eds. Alexei Miller and Maria Lipman (Budapest: Central European University Press, 2012): 279–308.

30. See Taylor Craig Zajicek, "Modern Friendship: The 'New Turkey' and Soviet Cultural Diplomacy, 1933–1934" (PhD dissertation, University of Washington, Seattle, 2014).

31. For discussion of Soviet five-year plans, see Donald J. Raleigh, *Provincial Landscapes: Local Dimensions of Soviet Power, 1917–1953* (Pittsburg: University of

Pittsburgh Press, 2001). For discussion of Turkish-Russian relationships (including Sergei Yutkevich's 1934 short film, *Ankara—Heart of Turkey*), Samuel J. Hirst, "Anti-Westernism on the European Periphery: The Meaning of Soviet-Turkish Convergence in the 1930s," *Slavic Review* 72, no. 1 (2013): 32–53. For discussion of Inönü policies, see Kıvanç Coş and Pinar Bilgin, "Stalin's Demands: Constructions of the 'Soviet Other' in Turkey's Foreign Policy, 1919–1945," *Foreign Policy Analysis* 6, no. 1 (2010): 43–60; John M. VanderLippe, *Politics of Turkish Democracy: Ismet Inonu and the Formation of the Multi-Party System, 1938–1950* (Albany: SUNY Press, 2012); and S. C. Wyatt, "Turkey: The Economic Situation and the Five-Years Plan," *International Affairs (Royal Institute of International Affairs 1931–1939)* 13, no. 6 (1934): 826–844. Note that Russia's urban and civil planning departments were updated in 1917 with the "Northern Russia hydro resources investigation team," and again in 1930 with the establishment of the Gidroproekt Institute; see Lagendijk, "Divided Development."

32. William Hale, "Ideology and Economic Development in Turkey 1930–1945," *British Journal of Middle Eastern Studies* 7, no. 2 (1980): 100–117. See also Zajicek, "Modern Friendship."

33. Sezgi Durgun, "Cultural Cold War at the Izmir International Fair: 1950s–60s," in *Turkey in the Cold War*, eds. Cangül Örnek and Çağdaş Üngör (New York: Palgrave Macmillan UK, 2013), 67–86.

34. Amiya Kumar Bagchi, *The Political Economy of Underdevelopment* (New York: Cambridge University Press, 1982), see p. 222; Zvi Yehuda Hershlag, *Introduction to the Modern Economic History of the Middle East* (Leiden: Brill, 1980), p. 194; John Kingsley Birge, *A Guide to Turkish Area Study* (Washington: Committee on Near Eastern Studies, American Council of Learned Societies, 1949). W. D. Hines and E. W. Kemmerer were American consultants for the Turkish five-year plan. For US contributions (e.g., those of General Electric) to Dnieper and other projects, see Wladimir Naleszkiewicz, "Technical Assistance of the American Enterprises to the Growth of the Soviet Union, 1929–1933," *The Russian Review* 25, no. 1 (1966): 54–76. And, for Russian visits to TVA, see Lagendijk, "Divided Development."

35. For Atatürk's Forest Farm, see Duygu Kacar, "A Unique Spatial Practice." For development of Ankara, see Burcu Yiğit-Turan, "Modernist Landscapes of Ankara," *Journal of Landscape Architecture* 11, no. 2 (2016): 14–25.

36. Cana Bilsel, "Ideology and Urbanism during the Early Republican Period: Two Master Plans for Izmir and Scenarios of Modernization," *Journal of the Faculty of Architecture* 16, no. 1–2 (1996): 13–60; Biray Kolluoğlu-Kırlı, "The Play of Memory, Counter-Memory: Building Izmir on Smyrna's Ashes," *New Perspectives on Turkey* 26 (2002): 1–28.

37. For background on Gorky Park, see Alexander Kalyukin, Thomas Borén, and Andrew Byerley, "The Second Generation of Post-Socialist Change: Gorky Park and Public Space in Moscow," *Urban Geography* 36, no. 5 (2015): 674–695; Claire Shaw, "A Fairground for 'Building the New Man': Gorky Park as a Site of Soviet Acculturation," *Urban History* 38, no. 02 (2011): 324–344.

38. Recent interest in the Izmir International Fair and Kültürpark includes Sezgi Durgun, "Cultural Cold War at the Izmir International Fair"; Biray Kolluoglu "Cityscapes and Modernity: Smyrna Morphing into Izmir," in *Ways to Modernity in Greece and Turkey: Encounters with Europe, 1850–1950* eds. Anna Frangoudaki and Caglar Keyder

(London: I. B. Tauris Publishers, 2007), 217–235; Vilma Hastaoglou-Martinidis, "Urban Aesthetics and National Identity: The Refashioning of Eastern Mediterranean Cities between 1900 and 1940," *Planning Perspectives* 26, no. 2 (2011): 153–182; and Jason Wood and B. Nilgün Öz, "The Turkish Amusement Park: Modernity, Identity and Cultural Change in the Early Republic," in *The Amusement Park: History, Culture and the Heritage of Pleasure*, ed. Jason Wood (London: Routledge, 2017), 98–119.

39. For discussion of the Russian businesses at the Izmir Fair, see Durgun, "Cultural Cold War at the Izmir International Fair." Russian machinery for agriculture in the 1930s is mentioned in the *Eastern European Papers, No. 5: Preliminary Report on Turkey, Part 1* (International Bank for Reconstruction and Development, March 1948).

40. Kortan, *Turkish Architecture and Urbanism through the Eyes of Le Corbusier*.

41. Bilsel, "Ideology and Urbanism," 26.

42. Cangül Örnek, "From Analysis to Policy: Turkish Studies in the 1950s and the Diplomacy of Ideas," *Middle Eastern Studies* 48, no. 6 (2012): 941–959.

43. Osman Demir and Yakup Emre Çoruhlu, "Determining the Property Ownership on Cadastral Works in Turkey," *Land Use Policy* 26, no. 1 (2009): 112–120, 113. See also H. Mustafa Palancıoğlu, Abdurrahman Geymen, and Mehmet Alkan, "The Turkish Cadastral Information System and Lessons Learned," *Scientific Research and Essays* 5, no. 7 (2010): 625–633. Areas of Turkey were mapped in analog format between 1924 and 1985; after this point digital approaches were used, a process that is ongoing and involves the General Directorate of Land Registry and Cadastre and the Ministry of Public Works and Settlement Land Registry. Official deeds are registered with the Cadastre Information System (Tapu ve Kadastro Bilgi Sistemi, TAKBİS).

44. Barın Kayaoğlu, "Strategic Imperatives, Democratic Rhetoric: The United States and Turkey, 1945–52: Cold War in the Aegean," *Cold War History* 9, no. 3 (2009): 321–345.

45. Craig Livingston, "'One Thousand Wings': The United States Air Force Group and the American Mission for Aid to Turkey, 1947–50," *Middle Eastern Studies* 30, no. 4 (1994): 778–825.

46. *Eastern European Papers, No. 5: Preliminary Report on Turkey, Part 1* (International Bank for Reconstruction and Development, March 1948).

47. R. Newton McDowele, President, Gary Slag Corporation, to US War Department, November 4, 1947, US National Archives (USNA), 867.154/9-1847, Microfilm, reel 1292, No. 11; R. Newton McDowele, President, Gary Slag Corporation, to Sumerbank, Ankara Turkey, November 4, 1947, USNA, 867.154/9-1847, Microfilm, reel 1292, No. 11.

48. Mr. Lovett from Mr. Wilds, January 23, 1948, USNA, 867.20/1-2348, Microfilm Real 1291, No. 11.

49. M. Martin to H. Sake, December 5, 1947, Turkish National Archives (TNA), Series 11, Folder 23, Box 5, File 30100; F. A. Barutçu to H. Sake, October 4, 1948 TNA, Series 14, Folder 700, Box 111, File 30100.

50. "About the Marshall Plan," TNA, Series 5, Folder 507, Box 80, File 30100.

51. *Eastern European Papers, No. 5: Preliminary Report on Turkey, Part 1* (International Bank for Reconstruction and Development, March 1948). See also the decision signed by then President İsmet Inönü on the establishment of American Credits

Commission, which aimed to manage the credits taken within the Marshall Plan. TNA, Series 10, Folder 48, Box 116, File 30182. On specifics for credit agreement with the European Development Bank for the Gediz, see TNA, Series 10, Folder 63, Box 199, File 30182.

52. *Milletlerarası İktisadi İşbirliği Teşkilatı* [International Organization for Economic Cooperation] 1952, 1–18. This report lists the distribution of funds among the different Turkish state departments between 1948 and 1951. TNA, Series 5, Folder 801, Box 125, File 3010. The Turkish Republic International Financial Cooperation Agency covers the initial period of 1951 and focused on specific departments (e.g., transportation, water, and agriculture). TNA, Series 3, Folder 807, Box 268, File 301000. For broad discussion of this topic, see Perin E. Gürel, *Limits of Westernization: A Cultural History of America in Turkey* (New York: Columbia University Press, 2017).

53. *Eastern European Papers No. 5: Preliminary Report on Turkey, Part 1* (International Bank for Reconstruction and Development, March 1948), 38.

54. US Embassy Ankara to US Department of State, September 18, 1950, USNA, NND 842934, Box 5430, 882.3312/9-1850.

55. Ibid.

56. Dorothy Thompson, "The Great Productive Possibilities of Turkey," *Daily Boston Globe*, June 1, 1949.

57. For discussion of this program in Çukurova (Adana plain), see Richard D. Robinson, "Tractors in the Village—A Study in Turkey," *Journal of Farm Economics* 34, no. 4 (1952): 451–462.

58. 18 September 1950. US Embassy Ankara to US Department of State, Washington. Farm Machinery (including tractors)—Turkey. RG59, 882.3312/9-1850, NA.

59. Murat Gül, *Emergence of Modern Istanbul: Transformation and Modernisation of a City* (London: I. B. Tauris, 2009).

60. City Planner's Views on Future Istanbul and Notes on Istanbul Municipal Improvement Program. Central File: Decimal File 883.021, Internal Economic, Industrial And Social Affairs, Housing. Rents. Building Construction. (Includes Commercial And Industrial Housing). Turkey. City And Country Planning. Elimination Of Congested And Slum Areas. "Blighted" Areas, September 20, 1956–May 10, 1957. September 20, 1956–May 10, 1957. MS Democracy in Turkey, 1950–1959: Records of the Department of State relating to Internal Affairs: Turkey, 1955–1959. National Archives (United States). Archives Unbound. Web. 21 Oct. 2016.

61. For Socony in Turkey, see Central File: Decimal File 883.2553, Internal Economic, Industrial And Social Affairs, Mines, Mining, Carbon, Graphite, Turkey, Petroleum. Oil, December 19, 1956–October 28, 1957. December 19, 1956–October 28, 1957. MS Democracy in Turkey, 1950–1959: Records of the Department of State relating to Internal Affairs: Turkey, 1955–1959. National Archives (United States). Archives Unbound. Web. October 21, 2016.

62. May 27, 1950. Ankara 398. Views of Fuat Koprulu. Central File: Decimal File 783.00, Internal Political And National Defense Affairs., Turkey, Political Affairs And Conditions. Elections. Political Parties And Groups. Political Refugees. Defectors. Amnesty. Revolutions, Riots, Disturbances. Civil War, May 27, 1950. 27 May 1950. MS Democracy in Turkey, 1950–1959: Records of the Department of State relating to Internal

Affairs: Turkey, 1950–1954. National Archives (United States). Archives Unbound. Web. 21 Oct. 2016.

63. May 27, 1950. Ankara 398. Views of Faut Koprulu.

64. Reorganization of the Court of Cassation. 14 January 1953. Central File: Decimal File 783.31, Internal Political And National Defense Affairs, Judicial Branch Of Government, Turkey, National Courts. Personnel. Rules, January 14, 1953. 14 Jan. 1953. MS Democracy in Turkey, 1950–1959: Records of the Department of State relating to Internal Affairs: Turkey, 1950–1954. National Archives (United States). Archives Unbound. Web. 21 Oct. 2016.

65. 17 September 1951. US Embassy Ankara to the Department of State, Washington. TRACTORS ON FARMS—TURKEY. Supplement to Ankara Dispatch #619, May 23, 1951, Farm Machinery. Unclassified. RG59, 882.3312/9-1751, NA. US unofficial numbers in 1949 tallied 5,000 tractors in the country of Turkey; in 1952 the number had increased to 25,000. Discussed in the April 3, 1953, Summary of the Electrification Program of Turkey. Central File: Decimal File 883.2614, Internal Economic, Industrial And Social Affairs, Engineering. Architecture, Public Works. Public Construction, Turkey, Public Utilities. Electricity. Water. Gas. Power Dams, March 3, 1950–December 4, 1954. March 3, 1950–December 4, 1954. MS Democracy in Turkey, 1950–1959: Records of the Department of State relating to Internal Affairs: Turkey, 1950–1954. National Archives (United States). Archives Unbound. Web. 23 Oct. 2016.

66. See, for example, 13 May 1952. US Embassy Ankara to the Department of State, Washington. Farm Machinery (including tractors)–Turkey. RG 59, 882.3312/5-1352, NA.

67. From Robinson, "Tractors in the Village," footnote 4.

68. Discussed in the April 3, 1953, Summary of the Electrification Program of Turkey. Central File: Decimal File 883.2614, Internal Economic, Industrial And Social Affairs. Engineering. Architecture, Public Works. Public Construction. Turkey. Public Utilities. Electricity. Water. Gas. Power Dams, March 3, 1950–December 4, 1954. March 3, 1950–December 4, 1954. MS Democracy in Turkey, 1950–1959: Records of the Department of State relating to Internal Affairs: Turkey, 1950–1954. National Archives (United States). Archives Unbound. Web. 23 Oct. 2016.

69. Mr. McGhee to Mr. Cummings, October 21, 1947, "Relationship between the Turkish Road Program and the Marshall Plan," 867.154/10-2147, NA, Microfilm 1292, No. 11. Memo dated January 30, 1948, New Attempts on the Water and Highway Problem. Statement by Mr. Kasim Gülek, Minister of Public Works: "We think of cooperating with the United States on our water works as we do on our road problem. We are in touch with the interested officials in American to invite American specialists who would help us on the matter." NA, Microfilm1292, No. 11.

70. Correspondences between members of TVA about (or with) Turkish citizens include the following: Dr. Herman Finer, Depart. of Gov. at Harvard University from Gordon R. Clapp, TVA General Manager, 8 Feb. 1945, Info Office 44-57 Sezer, TVA. Miss Marguerite Owen, Washington Rep. TVA from Gordon P. Merriam, Chief of Near Eastern Affairs, US Department of State, 21 December 1944, Info Office 44-57 Sezer, TVA. Gordon R. Clapp, TVA General Manager to Mukdum Osmay, Chief of the Agricultural Service, Turkish International Labour Office, 27 October 1944, Info Office 44-57 Sezer, TVA.

71. For discussion of the Bureau and Savage, see Chris Sneddon, *Concrete Revolution: Large Dams, Cold War Geopolitics, and the US Bureau of Reclamation*

(Chicago: University of Chicago Press, 2015). See also Savage to Blee, June 2, 1949. TVA. Demirel graduated in 1949 and then made his trip to the US Bureau of Reclamation. It is likely that he and Savage met during Savage's visit in Ankara in May and June 1949.

72. See correspondences between theTurkish Ambassador to the United States, Feridun G. Erkin, and Chairman of the TVA board, Gordon R. Clapp, to arrange for their visit to TVA in October 1952. Folder: Information Office 44-57. TVA. For Turkish Educators it was the University of Florida at Gainesville with the connection. See K. V. Wofford to W. A. Shafer, March 30, 1952, Information Office 44-57, TVA.

73. G. R. Clapp, H. C. Gökçen, E. Labonne, and D. Morton, Final Report—United Nations Economic Survey Mission for the Middle East—Part I (1949).

74. H. C. Gökçen to G. R. Clapp, December 13, 1950, Information Office 44-57, TVA.

75. N. Bass to V. C. Hare, April 23, 1952, Information Office 44-57, TVA.

76. H. J. Mathes to L. L. Scranton, December 10, 1956, Information Office 44-57, TVA.

77. Interview with Ambassador Gilbert A. Robinson, August 15, 2002, Association for Diplomatic Studies and Training Foreign Affairs Oral History Project.

78. Mr. Hickerson to Warren Kelchner, May 16, 1950. US Participation in the Izmir International Fair, August 20–September 20, 1950. Decimal File 883.191, Internal Economic. Industrial And Social Affairs. Industrial Matters. Nationalization. Automation. Turkey, Expositions. Exhibitions. Fairs, February 11, 1950–December 31, 1954. February 11, 1950–December 31, 1954. MS Democracy in Turkey, 1950–1959: Records of the Department of State relating to Internal Affairs: Turkey, 1950–1954. National Archives (United States). Archives Unbound. Web. 22 Oct. 2016.

79. H. Alpay Er, Fatma Korkut, and Özlem Er, "US Involvement in the Development of Design in the Periphery: The Case History of Industrial Design Education in Turkey, 1950s–1970s." *Design Issues* 19, no. 2 (2003): 17–34.

80. Interview with Ambassador Gilbert A. Robinson. Tensions at the Fair are also confirmed by Wilber Chase, Interview with Wilbur P. Chase, by Charles Stuart Kennedy, July 24, 1990. The Foreign Affairs Oral History Collection of the Association for Diplomatic Studies and Training. Chase gives details of the burning of the Greek pavilion. For Soviet and US actions at the Fair, see Durgun, "Cultural Cold War at the Izmir International Fair."

81. From the US Embassy, Ankara to Secretary of State in Washington, 27 January 1955, on the Izmir International Fair. Decimal File 883.191, Internal Economic, Industrial And Social Affairs. Industrial Matters. Nationalization. Automation. Turkey, Expositions. Exhibitions. Fairs. January 27, 1955–July 15, 1957. January 27, 1955–July 15, 1957. MS Democracy in Turkey, 1950–1959: Records of the Department of State relating to Internal Affairs: Turkey, 1955–1959. National Archives (United States). Archives Unbound. Web. 22 Oct. 2016. See also 26 February 1955. US Embassy Ankara to US Secretary of State, Washington. RG59, 882.191-1Z/2-2655, NA. 7 March 1960. C.E. Wells to USIA Office, Washington. Annual Report on US Exhibit Participation in International Events–Agricultural Exhibit at Izmir Trade Fair. RG59, 882.19/1-2761, NA.

82. Örnek, "From Analysis to Policy," 954.

83. 7 March 1960. C.E. Wells to USIA Office, Washington. Annual Report on US Exhibit Participation in International Events–Agricultural Exhibit at Izmir Trade Fair. RG59, 882.19/1-2761, NA. 23 August 1960. K.A. Byrns to US State Department.

Opening of Izmir Fair. RG 59, 882.19/1-2761, NA. August 31, 1962. American Consul Izmir to US State Department. 31st Izmir International Fair. RG59, 882.19/1-2761, NA. January 22, 1963. US Embassy Ankara to US State Department. Izmir International Trade Fair. RG 59, 882.19/1-2761, NA.

84. There is a much earlier history for water management during the Ottoman Empire, which included oversight by private companies, foundations (waqfs), and the central authorities. In 1826 the *Evkaf Nezareti* was established and brought greater centralization to foundations, including those engaged in water systems. See Hilmi Erdoğan Yayla, "Operating regimes of the government: Accounting and accountability changes in the Sultan Süleyman Waqf of the Ottoman Empire (The 1826 Experience)," *Accounting History* 16, no. 1 (2011): 5–34. For specifics on foundations and irrigation systems in the Gediz Valley, see Yūzō Nagata, *Tarihte âyânlar* [History of the Ayans]. Vol. 176 (Türk Tarih Kurumu Basımevi, Istanbul 1997).

85. Servet Özağaç, "Cumhuriyet Dönemi Türk Haritacılık Tarihi [The History of Turkish Cartography in the Republican Era]," *Yüksek Lisans Tezi, Ankara: TC Ankara Üniversitesi Türk İnkılâp Tarihi Enstitüsü* [Master's Thesis, Ankara: The Turkish Republic (TC) Ankara University Institute of Turkish Revolution History], 2006.

86. For an indication of the degree of impact from health-related issues (malaria) as well as water management planning schemes in the Izmir and Manisa regions, see "Foça v Menemen'de sıtma mücadelesine başlandı [The fight against malaria has begun in Foça and Menemen]," *Cumhuriyet*, August 2, 1930; "Nafia Vekaletinden [From the Ministry of Public Works]," *Cumhuriyet*, January 3, 1932; and a call specific to Lake Marmara to expand the lake into a reservoir by diverting water from the Gediz River by canal, "Mecliste dün Nafia ve Iktısat Vekaleti bütçeleri müzakere edildi [Budgets of Public Works and Economy Ministries negotiated yesterday in parliament]," *Cumhuriyet*, May 19, 1933.

87. Examples include "Nafia Vekaletinden," *Cumhuriyet*, January 12, 1930; "Nafia Vekaletinden," *Cumhuriyet*, March 2, 1930; "Nafia Vekaletinden," *Cumhuriyet*, January 11, 1932; and "From the Ministry of Public Works," *Cumhuriyet*, January 15, 1932.

88. News postings refer to a Turkish-Russian partnership for water management in Ödemiş: *Cumhuriyet*, Ekim 2, 1932; for infrastructure, *Cumhuriyet*, August 3, 1935; and more tender calls were placed in 1937 and 1938: "Nafia Vekaletinden," *Cumhuriyet* July 22, July 27, August 1, and August 6, 1937; "Nafia Vekaletinden," *Cumhuriyet*, June 1, 1938, with this posting run on June 9, June 17, June 30, July 5, July 12, and November 23, 1938. During the planning phases for water works in the Gediz Valley, consultants visited from Switzerland, the US, and Russia; see "Demirköprü Barajı ve Hidroelektrik Tesisleri Inşaatı [Construction of the Demirköprü Dam and Hydroelectric Plant]," *Türkiye Mühendislik Haberleri* (Eylül, 1955), 4–10; "Gediz' de Demirköprü Bendi ve Hidroelektik Santralı" [The Demirköprü Dam and Hydroelectric Station in the Gediz]," *Türkiye Mühendislik Haberleri* (Hazıran, 1955), 10–11; and "Demirköprü Barajı ve Mütfeferri Tesisleri"[The Demirköprü Dam and Related Facilities], *Türkiye Mühendislik Haberleri* (Kasım, 1957), 21–29.

89. Sabihia Gökçen flew over the area on April 4, 1938. She was Atatürk's adopted daughter.

90. "Izmirde [In Izmir]," *Cumhuriyet*, February 20, 1942; "Parti Grupu dün toplandı: Gediz nehrinin taşkınları hakkında verilen takrir görüşüldü [The party group met yesterday: Motion on Gediz floods discussed]," *Cumhuriyet*, May 10,

1944; "Egede su baskını [Inundation in the Aegean Region]," *Cumhuriyet*, January 21, 1948); "Anadoluda bir çok yerleri su bastı: Menderes, Gediz nehirleri taştı, bir çok köyleri seller kapladı [Many areas in Anatolia inundated: Menderes and Gediz Rivers overflow, floods cover many villages]," *Cumhuriyet*, February 13, 1948; "Su işlerinin tanzimi, Hükümet, 165 milyonluk ek tahsisat istiyor [Government reorganization of water works requires additional appropriation of 165 million]," *Cumhuriyet*, June 1, 1948; "GEDİZ Hidro-elektrik tesisatı yaptırılcaktır [Gediz Hydroelectric facility to be built]," *Cumhuriyet*, January 3, 1949; "Jeoloji Enstitüsü talebeleri tetkik seyahatine çıktılar [Geology Institute students set out on an investigative expedition]," *Cumhuriyet*, May 25, 1949; "Yale Üniversitesi talebelerinden bir grup, memleketimize gelecek [A group of students from Yale University will come to our country]," *Cumhuriyet*, May 25, 1949.

91. "About the Seyhan, Demirköprü, Kemer, and Hırfanlı Dams," Series 11, Folder 587, Box 94, File 30100, TNA.

92. For an overview of mid-twentieth-century water diplomacy wars in the Middle East, see Feliks Bochenski and William Diamond, "TVA's in the Middle East," *The Middle East Journal* (1950): 52–82, quote on page 55; Arthur Z. Gardiner, "Point Four and the Arab World: An American View," *The Middle East Journal* (1950): 296–306; John F. Kolars and William A. Mitchell, *The Euphrates River and the Southeast Anatolia Development Project* (Carbondale: SIU Press, 1991); Douglas Little, "Cold War and Covert Action: The United States and Syria, 1945–1958," *Middle East Journal* 44, no. 1 (1990): 51–75; and Paul Narziss Rosenstein-Rodan, "International Aid for Underdeveloped Countries," *Review of Economics and Statistics* 43, no. 2 (1961): 107–138.

93. To my knowledge, two public copies exist: the US Library of Congress and Indiana University. An additional copy is in the private library of the Karaosmanoğlu's Bağırgan estate.

94. KTA, *Gediz Basin Report* 1950 (III-24).

95. Appendix 6 to Chapter III of the KTA report, entitled "Local Interest in Industrial Expansion," includes a summary from questionnaires: (KTA 45–47).

96. 23 April 1950. Izmir Municipal Electric Power Plant and Facilities. Central File: Decimal File 883.2614, Internal Economic, Industrial And Social Affairs, Engineering. Architecture, Public Works. Public Construction, Turkey, Public Utilities. Electricity. Water. Gas. Power Dams. March 3, 1950–December 4, 1954. March 3, 1950–December 4, 1954. MS Democracy in Turkey, 1950–1959: Records of the Department of State relating to Internal Affairs: Turkey, 1950–1954. National Archives (United States). Archives Unbound. Web. 23 Oct. 2016.

97. KTA Gediz Basin Report 1950, Chapter III, pp. 111–144.

98. KTA Gediz Basin Report 1950: Chapter IX, page 13.

99. A few representative news articles can be found here: "World Bank Staff Ousted by Turkey," *New York Times*, March 21, 1954; "Turkey Demands World Bank Withdraw Its Representative," *Wall Street Journal*, March 22, 1954; "Recall Agent, Turkey Asks 'Interfering' World Bank," *Washington Post and Times Herald* (March 22, 1954).

100. "West Turkey Begins New Power Project," *New York Times*, March 25, 1955.

101. To Mr. J. Hampton Barnes, Eisenhower Exchange Fellowships, Inc., from Paul L. Evans, Director of Information, January 24, 1955, TVA.

102. For discussion of trainings, technical assistance, and water development, see Christopher W. Ivusic, "Training Experts of Other Nations," *Reclamation ERA* 54, no. 4 (1968): 92.

103. See Burcak Keskin Kozat, *Negotiating Modernization through US Foreign Assistance: Turkey's Marshall Plan (1948–1952) Re-interpreted* (PhD dissertation [Sociology], University of Michigan, 2006).

104. John F. Kolars and William A. Mitchell, *The Euphrates River and the Southeast Anatolia Development Project* (Carbondale: SIU Press, 1991). See page 27 for reference to Demirel.

105. J. H. Stratton to A. Menderes, April 10, 1956. Series 8, Folder 6, Box 6, File 30100, TNA.

106. Olivier Bouquet, "Old Elites in a New Republic: The Reconversion of Ottoman Bureaucratic Families in Turkey (1909–1939)," *Comparative Studies of South Asia, Africa and the Middle East* 31, no. 3 (2011): 588–600.

107. Nilgun Ü. Gülmez. "Ege Taşrasında 'Modern' in Mitleşmesi: Kemer (Bozdoğan, Aydın) Ve Demirköprü (Salihli, Manisa) Baraj Siteleri Ve Lojmanları [The Mythologisation of the 'The Modern' in the Rural Aegean: The Case of Kemer (Bozdoğan, Aydın, and Demirköprü (Salihli, Manisa) Dam Sites and Lodgings]," *METU Journal of the Faculty of Architecture* 32, no. 2 (2016): 19–43.

108. "Flood Perils Turkish Towns," *New York Times*, February 22, 1958.

109. For this suggestion, see "Demirköprü Barajı açıldı [The Demirköprü Dam opened]," *Milliyet*, May 23, 1960.

110. Er, Korkut, and Er, "US Involvement in the Development of Design in the Periphery."

111. See also Jo-Anne Pemberton, "The Changing Shape of Intellectual Cooperation: From the League of Nations to UNESCO," *Australian Journal of Politics & History* 58, no. 1 (2012): 34–50.

Chapter 4

1. Pertaining to what may have been said on September 4, 1963, I've checked the following archives: US Congressional Reports, US National Archives, JFK Archives, and the Smithsonian Archives. The details are perhaps somewhere, and hopefully they will be fully revealed one day. As described in this chapter, the only record of this event is from correspondences among archaeologists.

2. Paul Betts, "The Warden of World Heritage: UNESCO and the Rescue of the Nubian Monuments," *Past & Present* 226, no. S10 (2015): 100–125; Fekri A. Hassan, "The Aswan High Dam and the International Rescue Nubia Campaign," *African Archaeological Review* 24, no. 3–4 (2007): 73–94; Lynn Meskell, "The Practice and Politics of Archaeology in Egypt." *Annals of the New York Academy of Sciences* 925, no. 1 (2000): 146–169. Meskell, Lynn *A Future in Ruins: UNESCO, World Heritage, and the Dream of Peace*. London: Oxford University Press, 2018.

3. For an overview of the history of the US Army Corps of Engineers see Robert P. Grathwol and Donita M. Moorhus, *Bricks, Sand, and Marble: US Army Corps of Engineers Construction in the Mediterranean and Middle East, 1947–1991*, Vol. 45 (Washington, DC: Center of Military History, Corps of Engineers, US Army, 2009).

4. I thank Fikret Yegül for pointing out the critical importance of Jerash in Detweiler's history. For more information, see A. H. Detweiler, "Some Early Jewish Architectural Vestiges from Jerash," *Bulletin of the American Schools of Oriental Research* 87 (1942): 10–17; Nelson Glueck, "Jerash in the Spring of 1933: Preliminary Report of the Joint Expedition of the School in Jerusalem and Yale University," *Bulletin of the American Schools of Oriental Research* 53 (1934): 2–13; Carl Hermann Kraeling, *Gerasa, City of the Decapolis: An Account Embodying the Record of a Joint Excavation Conducted by Yale University and the British School of Archaeology in Jerusalem (1928–1930), and Yale University and the American Schools of Oriental Research (1930–1931, 1933–1934)* (Boston: American Schools of Oriental Research, 1938); and W. F. Stinespring, "Jerash in the Spring of 1934," *Bulletin of the American Schools of Oriental Research* 57 (1935): 3–9.

5. A. Henry Detweiler, *Manual of Archaeological Surveying*, Vol. 2 (New Haven, CT: American Schools of Oriental Research, 1948).

6. See A. Henry Detweiler's papers archived at Cornell University, AHD. CU. No finding aid currently exists for this collection.

7. Larry J. Hackman, interview with Lucius D. Battle, August 27, 1968; Lucius D. Battle, Oral History Interview—JFK#2, 8/27/1968. Papers of John F. Kennedy. Presidential Papers. President's Office Files. Larry J. Hackman interview with Lucius D. Battle, October 31, 1968; Lucius D. Battle, Oral History Interview—JFK#3, 10/31/1968. Papers of John F. Kennedy. Presidential Papers. President's Office Files.

8. Caquot was a member of the Academy of Sciences and Chairman of the French Committee. See his Soil Mechanics Report. AHD. CU.

9. Miles Glendinning, *The Conservation Movement: A History of Architectural Preservation: Antiquity to Modernity* (New York: Routledge, 2013).

10. Jack L. Davis and Natalia Vogeikoff-Brogan, "Introduction: Philhellenism, Philanthropy, or Political Convenience?" *Hesperia: The Journal of the American School of Classical Studies at Athens* 82, no. 1 (2013): 1–14; Despina Lalaki, "On the Social Construction of Hellenism Cold War Narratives of Modernity, Development and Democracy for Greece," *Journal of Historical Sociology* 25, no. 4 (2012): 552–577.

11. For overview of PL 480 see Robert H. Deans, "US Foreign Assistance Programs: The Impact of Local Currencies in Economic Development," *Journal of Developing Areas* 5, no. 4 (1971): 589–604; Willard W. Cochrane, "Public Law 480 and Related Programs," *ANNALS of the American Academy of Political and Social Science* 331, no. 1 (1960): 14–19. For specifics on archaeology, see Christina Luke and Morag Kersel, *US Cultural Diplomacy and Archaeology: Soft Power, Hard Heritage* (New York: Routledge, 2013), 21–35. Turkey was cut as a program country in 1963 and thus funds for archaeology were not available. The eligible countries were Egypt, Israel, Poland, Yugoslavia, Tunisia, India, Ceylon, Pakistan, Burma, and Guinea; see "The Smithsonian Institution Foreign Currency Program in Archeology and Related Disciplines," *American Journal of Archaeology*, 69, no. 4: 365.

12. Salem Nasr Habib, "Methods of Capturing the Potential Benefits of the Aswan High Dam in Egypt, UAR" (PhD dissertation, Georgia Institute of Technology, 1961); Ahmad Shokr, "Hydropolitics, Economy, and the Aswan High Dam in Mid-Century Egypt," *The Arab Studies Journal* 17, no. 1 (2009): 9–31; Amy L. S. Staples, "Seeing Diplomacy through Banker's Eyes: The World Bank, the Anglo-Iranian Oil Crisis, and the Aswan High Dam," *Diplomatic History* 26, no. 3 (2002): 397–418.

13. James E. Dougherty, "The Aswan Decision in Perspective," *Political Science Quarterly* 74, no. 1 (1959): 21–45.

14. Janos Horvath, "Economic Aid Flow from the USSR: A Recount of the First Fifteen Years," *Slavic Review* 29, no. 4 (1970): 613–632.

15. John F. Kennedy. Remarks at Muscle Shoals, Alabama, at the 30th Anniversary Celebration of TVA. May 18, 1963. Available from the American Presidency Project, http://www.presidency.ucsb.edu/ws/index.php?pid=9219.

16. B. M. Wedge to JFK, September 23, 1963. Papers of John F. Kennedy. Presidential Papers. President's Office Files.

17. Robert Rossow, "The Professionalization of the New Diplomacy," *World Politics* 14, no. 04 (1962): 561–575.

18. J. A. Wilson to JFK. Papers of John F. Kennedy. Presidential Papers. President's Office Files. Digital Identifier: JFKPOF-104-006.

19. See Timothy Mitchell, *Rule of Experts: Egypt, Techno-Politics, Modernity*. (Los Angeles: University of California Press, 2002).

20. For discussion of UNESCO and ICOMOS, see AHD. Box 20, C. 1.CU. See Mitchell's *Rules of Experts* for discussion of the challenges that the region faced due to the Aswan Project. See also Betts, "The Warden of World Heritage"; John Otis Brew, "Emergency Archaeology: Salvage in Advance of Technological Progress," *Proceedings of the American Philosophical Society* 105, no. 1 (1961): 1–10.

21. Memorandum from D. Goodwin to President J.F.K., January 11, 1863. Jacqueline Bouvier Kennedy Onassis Personal Papers. Series 1.1.4. Papers of John F. Kennedy. Presidential Papers. President's Office Files.

22. Betts, "The Warden of World Heritage"; Brew, John Otis. "Emergency Archaeology."

23. For discussion of class, race, and foreign diplomacy in Egypt, see Melani McAlister, "'The Common Heritage of Mankind': Race, Nation, and Masculinity in the King Tut Exhibit," *Representations* 54 (1996): 80–103.

24. Larry J. Hackman, interview with Lucius D. Battle, August 27, 1968; Lucius D. Battle, Oral History Interview—JFK#2, 8/27/1968. Papers of John F. Kennedy. Presidential Papers. President of the Institute of Public Administration. Mr. Luther Gulick, too, took interest and wrote to Mrs. Kennedy; see https://blogs.baruch.cuny.edu/ipaprocessing/2015/03/dear-jackie-save-abu-simbel/. See also the extensive correspondence between the First Lady and various figures in Washington regarding the gift of the Temple of Dendur to the United States. Jacqueline Bouvier Kennedy Onassis Personal Papers. Series 1.1.4. Papers of John F. Kennedy. Presidential Papers. President's Office Files. Digital Identifier: JBKOPP-SF043-001.

25. Larry J. Hackman, interview with Lucius D. Battle, October 31, 1968.

26. David Gissen, "The Architectural Production of Nature, Dendur/New York," *Grey Room* 34 (2009): 58–79. See also Douglas Little, "The New Frontier on the Nile: JFK, Nasser, and Arab Nationalism," *Journal of American History* 75, no. 2 (1988): 501–527.

27. For general discussion of PL 480, see Kenneth M. Kaufman and Helena Stalson, "US Assistance to Less Developed Countries, 1956–65," *Foreign Affairs* 45, no. 4 (1967): 715–725.

28. Larry J. Hackman, interview with Lucius D. Battle, October 31, 1968.

29. Ibid.

30. American Anthropological Association, "Smithsonian Institution Foreign Currency Program in Archeology and Related Disciplines," *Anthropology News* 6, no. 8 (1965): 4–5.

31. R. W. Komer to The President, January 9, 1963. Papers of John F. Kennedy. Presidential Papers. President's Office Files. Digital Identifier: JFKPOF-125-005. See also L. Carl Brown, ed., *Diplomacy in the Middle East: The International Relations of Regional and Outside Powers* (New York: I. B. Tauris, 2003); Nur Bilge Criss, "Strategic Nuclear Missiles in Turkey: The Jupiter Affair, 1959–1963," *Journal of Strategic Studies* 20, no. 3 (1997): 97–122; Süleyman Seydi, "Turkish–American Relations and the Cuban Missile Crisis, 1957–63," *Middle Eastern Studies* 46, no. 3 (2010): 433–455.

32. Ismet Inönü to JFK, October 23, 1962. Papers of John F. Kennedy. Presidential Papers. President's Office Files, 125-005.

33. R. W. Komer to The President (JFK), January 9, 1963. Papers of John F. Kennedy. Presidential Papers. President's Office Files, 125-005.

34. See George Maxim Anossov Hanfmann and Ilse Hanfmann, *Letters from Sardis* (Cambridge, MA: Harvard University Press, 1972); George M. A. Hanfmann, William E. Mierse, and Clive Foss, *Sardis from Prehistoric to Roman Times: Results of the Archaeological Exploration of Sardis, 1958–1975* (Cambridge, MA: Harvard University Press, 1983).

35. Cornell University President Deane W. Malott from A. Henry Detweiler, May 30, 1960. American Schools of Oriental Research (ASOR).

36. Nathan M. Pusey to Mr. Robert H. Thayer, Special Assistant to the Secretary of State, Department of State, June 6, 1960; an identical letter was sent to the Honorable Fletcher Warren American Ambassador, American Embassy, Turkey. ASOR.

37. A. Henry Detweiler to Ambassador Warren, June 20, 1960. ASOR.

38. Myron B. Smith, "American Institutes for the Near East," April 7, 1958, memorandum of conversation with Keith E. Adamson, officer at US Consulate General, Istanbul, Turkey. See also Donald Newton Wilber, *Adventures in the Middle East: Excursions and Incursions* (Princeton: Darwin Press, 1986).

39. Sardis Newsletter, "Report of the Field Director on His Trip to Turkey, March 20–April 14," dated April 12, 1958, from Istanbul. A. Henry Detweiler Papers (AHD). Cornell University (CU).

40. Senator George Aiken from Vermont served from 1953 to 1955.

41. Edward L. Schapsmeier and Frederick H. Schapsmeier, "Farm Policy from FDR to Eisenhower: Southern Democrats and the Politics of Agriculture," *Agricultural History* 53, no. 1 (1979): 352–371.

42. Feridun Cemil Özcan, "US Aid and Turkish Macroeconomic Policy: A Narration of the Aid Bargain Process in the 1946–1958 Period," *Turkish Yearbook of International Relations* 34 (2003): 119–136.

43. G. Hanfmann and I. Hanfmann, "Letters from Sardis," 122. See also the Interim Report on the Program, submitted to the Bureau of Educational and Cultural Affairs, Department of State, May 31, 1963. Discussion includes tourism planning and infrastructure with the Manisa Museum and local municipality. They also documented the visits from regional and national groups, academics, K-12 groups, and diplomats.

44. See American-Turkish Society, Inc., Newsletter, Nov. 1958, p. 2.

45. See *Türk Arkeoloji Dergisi* XII, No. 1 (1962): 26.

46. George Hanfmann to Turkish General Directorate, June 30, 1962, The 1961 Season. American Exploration of Sardis (AES), Harvard University (HU).

47. Willimar W. Warner, Special Assistant to the Secretary for International Activities at the Smithsonian to Altemur Kilic, Counselor for Information, Embassy of the Republic of Turkey, May 3, 1965. GUH. 65.15 (FP). Box 15 of 16. Correspondence and Other Papers, ca. 1945-1984. WUA. HU.

48. President Deane W. Malott from A. Henry Detweiler, April 15, 1962. ASOR.

49. G. Hanfmann, *Sardis from Prehistoric to Roman Times*, xi.

50. Fikret K. Yegül with contributions by Mehmet C. Bolgil, and Clive Foss, *The Bath-Gymnasium Complex at Sardis*, Vol. 3 (Cambridge, MA: Harvard University Press, 1986):152; Fikret K. Yegül, "The Marble Court of Sardis and Historical Reconstruction," *Journal of Field Archaeology* 3, no. 2 (1976): 169–194.

51. Yegül, "The Marble Court": 172.

52. The Sardis archives demonstrate the planning that took place. See also the references to this project woven into the various entries of G. Hanfmann and I. Hanfmann, "Letters from Sardis." Most informative are the detailed entries of Yegül's analyses (35, 36A, 36B, 37, 38) in Crawford Hallock Greenewalt, ed., *The City of Sardis: Approaches in Graphic Recording* (Cambridge: Harvard Art Museum, 2003).

53. Yegül, "The Marble Court," 173.

54. Ibid., 193.

55. Ibid., 174

56. G. Hanfmann Mierse, and Foss, *Sardis from Prehistoric to Roman Times*, 152.

57. G. Hanfmann and I. Hanfmann, *Letters from Sardis*, 198.

58. Kraeling, *Gerasa, City of the Decapolis*.

59. G. Hanfmann, *Sardis from Prehistoric to Roman Times,* xi.

60. Ibid.

61. G. Hanfmann and I. Hanfmann, *Letters from Sardis*, 13.

62. Jotham Johnson to George Hanfmann, February 21, 1964. HUG (FP) 65.15. Box 15 of 16. Correspondence and Other Papers, ca. 1945–1984. WUA. HU.

63. Jotham Johnson to George Hanfmann, March 2, 1964. HUG (FP) 65.15. Box 15 of 16. Correspondence and Other Papers, ca. 1945-1984. WUA. HU.

64. Jotham Johnson to George Hanfmann, March 2, 1964. HUG (FP) 65.15. Box 15 of 16. Correspondence and Other Papers, ca. 1945–1984. WUA. HU.

65. George Hanfmann to Jotham Johnson, February 23, 1964. HUG (FP) 65.15. Box 15 of 16. Correspondence and Other Papers, ca. 1945–1984. WUA. HU.

66. George Hanfmann to Ernst Kitzinger at Dumbarton Oaks, February 21, 1964. HUG (FP) 65.15. Box 15 of 16. Correspondence and Other Papers, ca. 1945–1984. WUA. HU.

67. Geroge Hanfmann to Luke Battle, April 22, 1964. Harvard Archives HUG (FP) 65.15. Box 15 of 16. Correspondence and Other Papers, ca. 1945–1984.WUA. HU.

68. William Warner to George Hanfmann, May 28, 1964. HUG 65.15. Correspondence and Other Papers, ca. 1945–1984. Box 15 of 16. WUA. HU.

69. Battle denied that he pursued Abu Simbel for the Ambassadorship.

70. Archaeological Exploration of Sardis. Final Report of the Program 1962–1966: Containing Comments on the Four Year Program and an Evaluation of Its Effectiveness. Submitted to the Bureau of Educational and Cultural Affairs, Department of State. See section, "Evaluation of the Program," pp. 31–32. AHD. CU.

71. George Hanfmann to William Warner, Feb. 26, 1965. A Proposal for the Continuation of the Program Known as Archaeological Exploration of Sardis. HUG (FP) 65.15. Box 15 of 16. Correspondence and Other Papers, ca. 1945–1984. WUA. HU.

72. George Hanfrmann to William Warner, February 26, 1965. HUG (FP) 65.15. Box 15 of 16. Correspondence and Other Papers, ca. 1945–1984. WUA. HU.

73. George Hanfmann to William Warner, February 26, 1965. A Proposal for the Continuation of the Program Known as Archaeological Exploration of Sardis. HUG (FP) 65.15. Box 15 of 16. Correspondence and Other Papers, ca. 1945–1984. WUA. HU.

74. These included work in Jordan (USAID under Detweiler with James Yarnell) and Peru (the Vicos program). Detweiler's engagement with USAID in Jordan is the topic of my ongoing research, notably the linkage with his early work at Jaresh; his ongoing work at Sardis; and then, his apparent re-entry to Jordan with Yarnell. The work of Cornell at Vicos has a very rich scholarly literature attached to it. The following recent essay provides a summary of that program: J. Pribilsky, "Development and the 'Indian Problem' in the Cold War Andes: Indigenismo, Science, and Modernization in the Making of the Cornell-Peru Project at Vicos," *Diplomatic History* 33, no. 3 (2009): 405–426.

Chapter 5

1. Hanfmann to Warner, February 26, 1965. HUG (FP) 65.15. Box 15 of 16. Correspondence and Other Papers, ca. 1945–1984. WUA. HU.

2. Alyattes, accessed May 27, 2017, http://alyattes.com/en/organic/.

3. For discussion of Bin Tepe see G. Hanfmann, *Sardis from Prehistoric to Roman Times: Results of the Archaeological Exploration of Sardis, 1958–1975* (Cambridge, MA: Harvard University Press, 1983). C. H. Roosevelt, *The Archaeology of Lydia, from Gyges to Alexander* (Cambridge, UK: Cambridge University Press, 2009).

4. In 1989 the slogan was changed to "Building Peace in the Minds of Men and Women."

5. Christopher H. Roosevelt and Christina Luke, "Central Lydia Archaeological Survey: 2007 Results," *Araştırma Sonuçları Toplantısı* 26, no. 2 (2009): 433–450; Christopher H. Roosevelt and Christina Luke, "Central Lydia Archaeological Survey: 2008 Results," *Araştırma Sonuçları Toplantısı* 27 (2010): 1–24.

6. See Christopher Ratté, "Not the Tomb of Gyges," *Journal of Hellenic Studies* 114 (1994): 157–161; Christopher Ratté, *Lydian Architecture: Ashlar Masonry Structures at Sardis.* Archaeological Exploration of Sardis Report 5 (Cambridge: Harvard University Press, 2011), Christopher H. Roosevelt, *The Archaeology of Lydia, from Gyges to Alexander* (Cambridge, UK: Cambridge University Press, 2009).

7. Christina Luke and Christopher H. Roosevelt, "Memory and Meaning in Bin Tepe, the Lydian Cemetery of the 'Thousand Mounds,'" *Tumulus as Sema: Space, Politics, Culture and Religion in the First Millennium BC* 27 (2016): 407–428.

8. See folder, RM Cook—George Dennis Excavations at AT. AES. HU. For general information on Dennis in Bin Tepe, see Butler, Howard C. *Sardis I. The Excavations, Part 1: 1910–1914.* (Leiden: E. J. Brill, 1922). Crawford H. Greenewalt, ed. *The City*

of Sardis: Approaches in Graphic Recording. (Cambridge, MA: Harvard University Art Museum, 2003).

9. Bekir Kambur, "Marmara Gölü Sulak Alanı'nın kültürel ekolojisi [Cultural Ecology of Lake Marmara's Wetlands]," (master's thesis, Social Science Institute, Balıkesir University).

10. Feridun Mustafa Emecen, *XVI. Asırda Manisa Kazası* [Manisa Province in the 16th Century]. (Ankara: Türk Tarih Kurumu Basımevi, 1989); Clive Foss, *Byzantine and Turkish Sardis*, No. 4 (Cambridge, MA: Harvard University Press, 1976).

11. This follows other patterns of Cossack/Kazak concessions for Lake Manyas; see Beglan B. Toğrol, "The Kuban Cossacks of Manyas Turkey and 'The Cossacks' of Leo Tolstoy," *Psikoloji Çalışmaları Dergisi* (1987): 125–146; and J. A. R. Munro, "Dascylium," *Journal of Hellenic Studies* 32 (1912): 57–67. On page 59, Munro refers to Hasluck's scholarship, F.W. Hasluck, "Unpublished Inscriptions from Cyzicus Neighbourhood," *Journal of Hellenic Studies* 24 (1904): 20–40. Hasluck refers to the Kazakkeui on the lake of Manyas on p. 22. Turkish sources include Ziyaeddin Fahri Fındıkoğlu, "Rus Kazakları Etrafında Nazariyeler ve Bir Ek [Theories about the Russian Cossacks and an Appendix]," *İktisat Fakültesi Mecmuası* 23, no. 3–4 (1963): 151–166. Specific mention of Salihli Kazaks is found in Mustafa Kalkan, "Sovyetler Döneminde Kazakların Göç Hareketleri ve Anadolu'da (Altay Köyü'nde) Iskan Edilşeli [Emigration Movements of the Kazakh Turks in the Soviet Era and their Settlement in Anatolia (the Village of Altay)]," TÜBAR-XXI (2007): 123–141. See E. Giraud, *Days Off*, p. 10, for mention of Circassian leadership of the fisheries. For discussion of the Kazak community in Salihli through the 1970s, see Orhan Türkdoğan, "Türkiye'de Göçler, Sorunları ve Çözüm Yolları [Migrations in Turkey: Problems and Solutions]," *İktisat Fakültesi Mecmuası* 35, no. 1–4 (1975): 40–67; Yılmaz, Arı, "Manyas Gölü'nün kültürel ekolojisi: Tarihi süreçte adaptasyon ve değişimi [The Cultural Ecology of Lake Manyas: Adaptation and Change in Historical Perspective]," *Türk Coğrafya Dergisi* 40 (2003): 75–97. For specifics on Crimea and Caucasus groups emigrating to Manisa, see Ferhat Berber, "Kırımlı ve Kafkasyalı Muhacirlerin Manisa Kazasında İskanı (1860–1876) [The Crimean and Caucasian Migrants in Manisa Province]," *Celal Bayar Üniversitesi Sosyal Bilimler Dergisi* 11, no. 3 (2013): 169–192.

12. Christina Luke, Christopher H. Roosevelt, and Catherine B. Scott, "Yörük Legacies: Space, Scent, and Sediment Geochemistry," *International Journal of Historical Archaeology* (2016): 1–26.

13. Yūzō Nagata, *Tarihte âyânlar* [History of the Ayans]. Vol. 176 (Turk Tarih Kurumu Basmevi, Istanbul 1997). Oral histories were collected as part of the Central Lydia Archaeological Survey (CLAS). Christina Luke and Elvan Cobb, "Dwelling in Hacıveliler: Social-Engineering Policies in the Context of Space, Place and Landscape in Rural, Western Turkey," *Anatolian Studies* 63 (2013): 155–173; Luke, Roosevelt, and Scott, "Yörük Legacies."

14. Nagata, *Tarihte âyânlar.*

15. Teoman Ergül, *Kurtuluş Savaşında Manisa, 1919–1922* [Manisa during the Turkish War of Independence, 1919–1922], Vol. 1 (Manisa: Manisa Kültür Sanat Kurumu, 1991). See also Christina Luke, "Archaeological Heritage and the Turkish War of Independence in Central Lydia," *Museum International* 62, no. 1-2 (2010): 42–47.

16. For overview of land policies in Turkey, see Osman Demir and Yakup Emre Çoruhlu, "Determining the Property Ownership on Cadastral Works in Turkey," *Land Use Policy* 26, no. 1 (2009): 112–120.

17. This concept is described by Bouquet, although not specifically for this estate. See Olivier Bouquet, "Old Elites in a New Republic: The Reconversion of Ottoman Bureaucratic Families in Turkey (1909–1939)," *Comparative Studies of South Asia, Africa and the Middle East* 31, no. 3 (2011): 588–600.

18. US Embassy in Ankara to US Department of State, Washington, DC. Turkish Historical and Archaeological Sites. June 23, 1950. 882-2613/6-2350. RG 58 Box 5430. NND842934. USNA.

19. Regarding maps and images, see Kamil Su, Director General Directorate of Antiquities and Museums to George Hanfmann, February 5, 1958. AES, Permits, 1958 Season; George Hanfmann to Rüstem Duyuran, Ministry of Education, March 5, 1962. AES. Permits, 1961 Season.

20. George M. A. Hanfmann, *Sardis from Prehistoric to Roman Times: Results of the Archaeological Exploration of Sardis, 1958–1975* (Cambridge, MA: Harvard University Press, 1983); David G. Mitten and G. Yüğrüm, "Excavations at Ahlatlı Tepecik on the Gygean Lake, 1968," *Türk Arkeoloji Dergisi* 17 (1968): 125–131. David G. Mitten and G. Yüğrüm, "The Gygean Lake, 1969: Eski Balıkhane, Preliminary Report," *Harvard Studies in Classical Philology* 75 (1971): 191–195; David G. Mitten and G. Yüğrüm, "Ahlatlı Tepecik beside the Gygean Lake," *Archaeology* 27 (1974): 22–29; Christopher H. Roosevelt, "Central Lydia Archaeological Survey: 2005 Results," *Araştırma Sonuçları Toplantısı* 24 (2007): 135–154.

21. George Hanfmann to William Warner, Special Assistant to the Secretary for International Activities at the Smithsonian, February 26, 1965. A proposal for the continuation of the program known as Archaeological Exploration of Sardis. HUG (FP) 65.15. Box 15 of 16. WUA. HU.

22. Christopher Ratté, "Not the Tomb of Gyges," *Journal of Hellenic Studies* 114 (1994): 157–161.

23. G. Hanfmann and I. Hanfmann, *Letters from Sardis*, 192–194. Extraction efforts to locate the tomb were renewed in 2013. The tunnels were stabilized, but again the mound won—no chamber (plundered or pristine) was located.

24. Sanka Knox, "Archaeologists Find Huge Tomb of Lydian King," *New York Times*, November 8, 1964.

25. See "Metropolitan Museum Queried by Turks on Smuggled Artifacts," *New York Times*, August 27, 1970; and "Turks Warn of Bar to Archaeologists If the US Won't Aid Smuggling Fight," *New York Times*, November 3, 1970. For details of this case, see İlknur Özgen, Jean Öztürk, and Machteld Johanna Mellink, *The Lydian Treasure: Heritage Recovered* (Uğur Okman for Republic of Turkey, Ministry of Culture, General Directorate of Monuments and Museums, 1996); and Christopher H. Roosevelt and Christina Luke, "Mysterious Shepherds and Hidden Treasures: The Culture of Looting in Lydia, Western Turkey," *Journal of Field Archaeology* 31, no. 2 (2006): 185–198.

26. George Hanfmann to William Warner, February 26, 1965. A proposal for the continuation of the program known as Archaeological Exploration of Sardis. HUG (FP) 65.15. Box 15 of 16. WUA. HU.

27. George Hanfmann to Director General of Antiquities and Museums, Ankara, August 29, 1974. HUG (FP). 65.7. Correspondence and Other Papers, ca. 1931–1985, Box 10 of 19. WUA. HU. George Hanfmann to Director General of Antiquities and Museums, September 19, 1974. HUG (FP). 65.7. Correspondence and Other Papers, ca. 1931–1985, Box 10 of 19. WUA. HU.

28. Turkish text: "a-) I. Derece Arkeolojik Sit alanarı olarak belirlenen höyük ve tümülüslerde toprağın sürülmesine dayanan tarımsal faaliyetlerin kesinlikle yasaklanmasına, ağaçlandırmaya gidilmemesine, yalnızca mevcut ağaçlardan ürün alınabileceğine; b -) Her türlü kazı yapmanın önlenmesine (bilimsel amaçlı alanların dışında) toprak, taş alınmanasına, kireç, teş, tuğla, mermer (v.s. gibi) ocakların açılmamasına; c -) Zorunlu durumlarda ilgili resmi ve özel kuruluşlarca dikilecek direkler, açılacak su kanalları, boru hatları, karayolları yapımı vb. uygulamalar için koruma kurulundan özel izin alınmasına; d -) Bu gibi alanlarda sınırlarının sınır taşları veya tel örgüler ile belirlenmesine tanıtıcı levhaların Kültür Bakanlığı tarafından konulmasına, bu alanlar için bekçi sağlanmasına yerel yönetimlerce gerekeli önlemlerin alınmasına karar verildi."

29. Unofficial translation of the Ministry of Culture, First Izmir Protection Council for the Preservation of Cultural and Natural Assets, #1.6.1990-1946, 1 Ocak 1990.

30. From 2005 to 2013 the Central Lydia Archaeological Survey conducted systematic, diachronic survey in the Marmara Lake Basin, including Bin Tepe proper.

31. Christina Luke, "Social Routes of Heritage," paper presented at the 40th World Heritage conference in Istanbul, 12 July 2016 Turkey.

32. Güllük Dagi-Termessos National Park (2000) is listed, but no specific criteria are given.

33. See also, Ramsar Sites of Turkey (Ankara: Doğa Derneği, 2011). Bolca, Mustafa, Fulsen Özen, and Aslı Güneş. "Land use changes in Gediz Delta (Turkey) and their negative impacts on wetland habitats." *Journal of Coastal Research* 30, no. 4 (2012): 756–764.

34. Asuman Abacioğlu, "Sulak alanlar kirlilik kıskacında,"[Wetlands in jeopardy from pollution] *Cumhuriyet*, February 5, 1995. See also (for a larger perspective) Annabelle Cuttelod, Nieves García, Dania Abdul Malak, Helen J. Temple, and Vineet Katariya, "The Mediterranean: A Biodiversity Hotspot under Threat," *Wildlife in a Changing World—An Analysis of the 2008 IUCN Red List of Threatened Species* 89 (2009): 89–101.

35. See Nilgun B. Harmancioglu, Kurt Fedra, and Filiz Barbaros, "Analysis for Sustainability in Management of Water Scarce Basins: The Case of the Gediz River Basin in Turkey," *Desalination* 226, no. 1–3 (2008): 175–182.

36. The two endangered endemic fish are the *Knipowitschia mermere* and the *Ladigesocypris mermere*; the most threatened birds include the Dalmatian pelican, ferruginous duck, white-headed duck, and the spur-winged lapwing or spur-winged plover. See "Marmara Lake," *Bird Life International*, accessed May 15, 2017, http://datazone.birdlife.org/site/factsheet/marmara-lake-iba-turkey.

37. Luke, Roosevelt, and Scott, "Yörük Legacies."

38. For a general summary of the project, yet not specific to the gardens, see Nilay Ünsal Gülmez, "Ege Taşrasında 'Modern' in Mitleşmesi: Kemer (Bozdoğan, Aydın) Ve Demirköprü (Salihli, Manisa) Baraj Siteleri Ve Lojmanları," *METU Journal of the Faculty of Architecture* 32, no. 2 (2016).

39. The relocation of Yeniköy, Kılcanlar, and Beyler from the mountains to the northern shores of Lake Marmara facilitated the transition to intensive agriculture. See Luke and Cobb, "Dwelling in Hacıveliler."

40. In the summer of 2017 the Marmara Regulator compound was in ruins. The Ahmetli compound was in disrepair, yet still in use.

41. A deed-book (smaller than all others around it) sits on the shelf in the archive room at the Izmir State Water Works office (DSİ). The files are specific to Bin Tepe and the shores of Lake Marmara. They show that a landholder was often deeded property in the morning, and, if the land was required for the hydroelectric and irrigation project, it was expropriated in the afternoon of the same day. See also Luke and Cobb, "Dwelling in Hacıveliler."

42. Provisions for expropriation follow the 1925 Public Expropriation Decree, and, more recently, Expropriation Law 2942, amended No. 4650 in 1983 (46th article of the Turkish Constitution). Similar to other countries, in Turkey compulsory expropriation of private properties may proceed without consent owner see Tahsin Boztoprak, Osman Demir, and Yakup Emre Çoruhlu, "Comparison of Expropriation and Land Consolidation on the Regulation of Agricultural Land," *Sigma* 34, no. 1 (2016): 43–55.

43. Gustave Burmeister, "The Current and Prospective Foreign Trade Situation with Respect to Farm Products," *Journal of Farm Economics* 36, no. 5 (1954): 973–989.

44. See Chapter 3 (this volume) for discussion of the first five-year plan, which was unveiled in 1934. But it was not fully implemented. The new model, specific to the 1963 plan, has continued to structure the Turkish economy, now in its tenth cycle.

45. David Gisselquist and Carl Pray, "The Impact of Turkey's 1980s Seed Regulatory Reform," in *Easing Barriers to Movement of Plant Varieties for Agricultural Development*, ed. D. Gisselquist and J. Srivastava (Washington, DC: Banco Mundial. 1997): 113–131.

46. Donald Larson, Will Martin, Sebnem Sahin, and Marinos Tsigas, "Agricultural Policies and Trade Paths in Turkey," *World Economy* (2016): 1194–1224.

47. Zülküf Aydin, "Neo-Liberal Transformation of Turkish Agriculture," *Journal of Agrarian Change* 10, no. 2 (2010): 149–187.

48. Alper Demirdöğen, Emine Olhan, and Jean-Paul Chavas, "Food vs. Fiber: An Analysis of Agricultural Support Policy in Turkey," *Food Policy* 61 (2016): 1–8.

49. Llewellyn Mealey Humphrey, *Mexican Wheat Comes to Turkey* (Food and Agricultural Division, United States Agency for International Development, Turkey, April 1969), 16.

50. *Meksika Buğdayı'nın yetiştirilmesi* [The Growing of Mexican Wheat]. (Tarim Bakanlığı: Buğday Bülteni No. 3, 1969); Coşkun Saydam, Mustafa Öğüt, and Mustafa Copcu, "Ege Bölgesinde Yetiştirlilen Meksika Kaynaklı Buğdayların Hastaklıklarla Ilgisi ve kuru tohum ilaçlamasının sürme gücüne etkisi üzerinde çalışmalar [Studies on Diseases of Mexican Wheat Grown in the Aegean Region and the Effect of Pesticides on Dry Seed Germination]," *Bitki Koruma Bülteni Cilt* 14, no. 3 (1974): 151–180.

51. Humphrey, *Mexican Wheat Comes to Turkey*, 24–26.

52. Louis P. Reitz, "New Wheats and Social Progress," *Science* 169, no. 3949 (1970): 954.

53. Humphrey, *Mexican Wheat Comes to Turkey*, 30–42.

54. Scholars report that many tractors were unable to be used due to the impossibility of securing spare parts. They were referred to as "the Marshall Mules." See Perin E. Gürel, *Limits of Westernization: A Cultural History of America in Turkey* (New York: Columbia University Press, 2017). In contrast, the region of Manisa had good access to the city and port of Izmir, and thus did not experience this problem.

55. Attila was the son of Ethem (Bey) Karaosmanoğlu, who had been a state official in railways. While he was not directly related to Lutfi of the Bağırgan, they were distant

relatives. See Attila Karaosmanoğlu, *Izmir Karşıyaka'dan dünya'ya* [From Izmir's Karşıyaka to the world]. Vol. 890. (Türkiye İş Bankası Kültür Yayınları, 2005).

56. *OECD 50th Anniversary Vision Statement*, accessed May 7, 2018, https://www.oecd.org/mcm/48064973.

57. Transcript of interview with Attila Karaosmanoglu, November 17, 1994, and January 10 and 18, 1995, World Bank Group Historian's Office, Oral History Program.

58. Attila served as Chief Economist in the Europe, Middle East, and North Africa Region from 1973 to 1975; Director in the Development Policy Vice Presidency from 1975 to 1979; Director in the Country Programs Department 1, Europe, Middle East, and North Africa Region from 1979 to 1982; Vice President in the East Asia and Pacific Region from 1983 to 1987; Vice President in the Asia Region from 1987 to 1991; and Managing Director (MDS) from 1991 to 1994. See World Bank Archives, accessed March 5, 2017, https://archivesholdings.worldbank.org/attila-karaosmanoglu-files.

59. Sophia Davidova and Kenneth Thomson, "Family Farming in Europe: Challenges and Prospects," in *Agriculture and Rural Development* (European Parliament Publication Office, 2014).

60. Mirela Tomas-Simin and Danica Glavas-Trbic, "Historical Development of Organic Production 1," *Ekonomika Poljoprivrede* 63, no. 3 (2016): 1083; Holger Kirchmann et al., "Fundamentals of Organic Agriculture—Past and Present," in *Organic Crop Production—Ambitions and Limitations* (Netherlands: Springer, 2008), 13–37.

61. K. Demiryürek, C. Stopes, and A. Güzel, "Organic Agriculture: The Case of Turkey," *Outlook on Agriculture* 37, no. 4 (2008): 261–267.

62. The organic sector is expanding in Turkey: currently there are farmer training and extension agents in eighty-one provinces. See Zerrin Kenanoğlu and Özlem Karahan, "Policy Implementations for Organic Agriculture in Turkey," *British Food Journal* 104, no. 3/4/5 (2002): 300–318; Orhan Özçatalbaş, "Current Status of Advisory and Extension Services for Organic Agriculture in Europe and Turkey," in *Organic Agriculture towards Sustainability*, ed. Vytautas Pilipavicius (London: InTechOpen, 2014), 1301–1375.

63. For summary of Rapunzel in Tekelioğlu, see http://www.rapunzel.de/uk/projekte.html, accessed on June 14, 2017.

64. M. M. Artukoglu, E. Tarkan, F. Gencler, and B. Miran, "Evaluating the Factors of Transition in Organic Cotton Production for Farmers: Case of Salihli, Turkey," *Bulgarian Journal of Agricultural Science* 15, no. 1 (2009): 77–83. In 1980, 90% of Turkey's cotton was grown in irrigated fields. See James L. Forsythe, "World Cotton Technology since World War II," *Agricultural History* 54, no. 1 (1980): 208–222.

65. Kesin İnşaat Yasağı getirilen korunması gerekli taşınmaz Kültür ve Tabiat Varlıklarının Bulunduğu Sit Alanlarındaki Taşınmaz Malların Hazineye Ait Taşınmaz Mallar ile Değiştirilmesi Hakkında Yönetmelik [Regulation on the Exchange of Real Estate Property Located in Protection Zones Where Required Protection of Immovable Cultural and Natural Heritage Results in Certain Construction Prohibitions with Real Estate Property Belonging to the Treasury], Official Gazette dated 2/2/1990, No. 20427, GM/0743E.050.

66. Decision of the 18th Civil Chamber of Court of Cassation, Case No: 1994/15418, Decision No. 1995/63. 1/16/1995. GM/3194 H.460.

67. Decision of the 1st Civil Chamber of Court of Cassation, Case No: 1995/7198, Decision No. 1995/5382, 11/4/1995, GM 3402 H.544; Decision of the General Assembly

of the Court of Cassation, Case No. 1995/211, Decision 1995/318, 12/4/1995, GM/0743 H.304; Decision of the 7th Civil Chamber of Court of Cassation, Case No: 1996/7055, Decision No. 1996/9414, 11/27/1996, GM/3402 H.631.

68. "Türk zeytinyağına Japonya'dan ödül Tokyo'da Japon Tadımcılar Birliği'nce 18–22 Nisan 2012 tarihleri arasında düzenlenen Olive Japan etkinliği çerçevesinde yapılan kalite yarışmasında Türk zeytinyağları büyük başarı gösterdi [Award from Japan to Turkish olive oil: Turkish olive oils showed great success in quality contest held in the framework of the Olive Japan event held between April 18–22, 2012 by the Japanese Tasters Union in Tokyo]," *Milliyet*, April 25, 2012.

69. The World of Organic Agriculture 2017 report by IFOAM EU Group: European Union Group of IFOAM—Organics International, accessed February 20, 2017, http://www.organic-world.net/yearbook/yearbook-2017/pdf.html.

70. In 2017, an interview with a prosperous olive farmer revealed that a farmer with a few acres of land planted with mature trees will likely clear 10,000 TL per year and upward, depending on the size of the property and number of trees. During the height of the olive season, a regional representative will process approximately 600 crates of table olives per evening, netting close to 14,000 TL per night. According to the Turkish Statistical Institute, the average annual household income in Turkey in 2016 was 22,732 TL. These numbers make it clear that olive orchards in Manisa may be very lucrative.

71. "Sit alanı tarım arazisi olacak, [Protected zoning to become agricultural land]," *Internet Haber*, Mayis 16, 2010); "Pazarköy'ün zeytinci Hanımağası [Pazarkoy's Olive Growers women's network]," *Salihli Vizyon*, Ekim-Kasim-Aralik 2011: 22–23; "Köyde herkes 'Ağır Ceza'lık ['Heavy Penalties' for Everyone in Villages]," *Turkiyeturizm.com*; "Bintepeler Tümüsleri'ndeki sit alanının yeniden belirlenmesi istendi [Revision of Protected Zoning for Tumuli in Bin Tepe Requested]," *Birgün, Manisa'nın İnternet Gazetesi*, Ekim 20, 2016).

Reflections

1. This is the work of the famous Italian sculptor Pietro Canonica, also known for his Taksim Republic Monument in the heart of Istanbul. Faik Gur, "Sculpting the Nation in Early Republican Turkey," *Historical Research* 86, no. 232 (2013): 342–372.

2. Yasemin Uzakgören, "Design Criteria for Increasing Environmental Performance of Recreational Areas: Analysis of Kordon/Izmir" (master's thesis, Izmir Institute of Technology, 2010).

3. For Istanbul on this topic, see Christina Luke, "The 40th World Heritage Session in Istanbul, Turkey: A Reflection on the Legacies of Heritage Policy and Missed Mega-Heritage," *Journal of Field Archaeology* 41, no. 6 (2016): 641–644.

4. Uğur Emek, "Turkish Experience with Public Private Partnerships in Infrastructure: Opportunities and Challenges," *Utilities Policy* 37 (2015): 120–129. For Central Asia and Turkey, see Ali M. Koknar "Turkey's security relations in Central Asia," in *Eurasia in Balance: The US and the Regional Power Shift*, ed. Ariel Cohen (New York: Routledge, 2017), 101–126.

5. For the Adala solar projects, see "Salihli-Adala Solar Power Plant Environmental Management Plan, Adala District, Salahli Manisa: Parcel 2250" (Ankara: PPM Pollution Prevention and Management Co. 2016). See also the projects under berakenerji: http://www.berakenerji.com/en/our-power-plants.aspx. And for wind projects, the energy

institute, accessed February 5, 2018, http://enerjienstitusu.com/2017/04/20/izmir-manisa-balikesir-ruzgar-enerjisiyle-3-milyon-evin-elektrik-ihtiyacini-karsiliyor/.

6. See Esra Özyürek, "Miniaturizing Atatürk Privatization of State Imagery and Ideology in Turkey," *American Ethnologist* 31, no. 3 (2004): 374–391. For a deeper history of the topic, see Faik Gur, "Sculpting the Nation in Early Republican Turkey," *Historical Research* 86, no. 232 (2013): 342–372; and Christopher S. Wilson, *Beyond Anitkabir: The Funerary Architecture of Atatürk: The Construction and Maintenance of National Memory* (London: Routledge, 2016).

7. In recent years, archaeological permits for these projects have been suspended. See Erik Stokstad, "Turkish Government Shuts Down Important Archaeological Dig, Apparently to Punish Austria," *Science*, September 6, 2016. In a display of great restraint, German archaeologists managed to keep Pergamon (and Hattuşa) out of the mainstream English media. This was not the case in 2013, when tensions over permits and repatriation between Turkey and Germany graced the pages of newspapers. See Christina Luke and Morag Kersel, *US Cultural Diplomacy and Archaeology: Soft Power, Hard Heritage* (London: Routledge, 2013), 44–46, 61–62.

8. See Özgen Acar, "Canal Project Threatens Ancient Ephesus in Western Turkey," *Hurriyet Daily News*, October 23, 2016, accessed February 8, 2018, http://www.hurriyetdailynews.com/civilizations-massacre-in-ephesus-121230; and see "Ambitious Project Aims to Reunite Ancient City of Ephesus with Sea," *Hurriyet Daily News*, September 26, 2017, accessed February 8, 2018, http://www.hurriyetdailynews.com/ambitious-project-aims-to-reunite-ancient-city-of-ephesus-with-sea-118416.

9. This issue reared its head during the final months of writing this book. See Izmir 2 Numarlı Kültür Varlıklarını Koruma Bögle Kurulu Karar No. 24.11.2017.9101.

10. Christopher J. Barrow, "River Basin Development Planning and Management: A Critical Review," *World Development* 26, no. 1 (1998): 171–186.

11. Lynn Meskell, "Human Rights and Heritage Ethics," *Anthropological Quarterly* 83, no. 4 (2010): 839–859. See also Meskell's work on heritage and development in South Africa: Lynn Meskell, "From Paris to Pontdrift: UNESCO Meetings, Mapungubwe and Mining," *The South African Archaeological Bulletin* (2011): 149–156; and in Panama: Lynn Meskell, "States of Conservation: Protection, Politics, and Pacting within UNESCO's World Heritage Committee," *Anthropological Quarterly* 87, no. 1 (2014): 217–243.

12. Rodney Harrison, "Beyond 'Natural' and 'Cultural' Heritage: Toward an Ontological Politics of Heritage in the Age of Anthropocene," *Heritage & Society* 8, no. 1 (2015): 24–42.

13. Will Steffen, Jacques Grinevald, Paul Crutzen, and John McNeill, "The Anthropocene: Conceptual and Historical Perspectives," *Philosophical Transactions of the Royal Society of London A: Mathematical, Physical and Engineering Sciences* 369, no. 1938 (2011): 842–867; Paul J. Lane, "Archaeology in the Age of the Anthropocene: A Critical Assessment of Its Scope and Societal Contributions," *Journal of Field Archaeology* 40, no. 5 (2015): 485–498; Kathryn Lafrenz Samuels, "Transnational Turns for Archaeological Heritage: From Conservation to Development, Governments to Governance," *Journal of Field Archaeology* 41, no. 3 (2016): 355–367.

14. Recent studies of this paradigm include Jussi P. Laine and Anna Casaglia, "Challenging Borders: A Critical Perspective on the Relationship between State,

Territory, Citizenship and Identity: Introduction," *Europa Regional* 24, no. 1/2 (2017): 3–8; and Richard Schofield, "International Boundaries and Borderlands in the Middle East: Balancing Context, Exceptionalism and Representation," *Geopolitics* (2017): 1–24.

15. Chris Sneddon and Coleen Fox, "The Cold War, the US Bureau of Reclamation, and the Technopolitics of River Basin Development, 1950–1970," *Political Geography* 30, no. 8 (2011): 450–460. Note that Sneddon and Fox do not make the explicit link to UNESCO, yet they do to Arab-TVA. See Chapter 3 (this volume) for UNESCO and Arab-TVA linkage.

16. Timothy Mitchell, *Rule of Experts: Egypt, Techno-Politics, Modernity* (University of California Press, Berekely 2002).

17. For discussions of heritage and statecraft, see Gertjan Plets, "Heritage Bureaucracies and the Modern Nation State: Towards an Ethnography of Archaeological Systems of Government," *Archaeological Dialogues* 23, no. 2 (2016): 193–213; and Gertjan Plets, "Heritage Statecraft: When Archaeological Heritage Meets Neoliberalism in Gazprom's Resource Colonies, Russia," *Journal of Field Archaeology* 41, no. 3 (2016): 368–383.

18. Diplomatic Note 00736. United States Embassy, Ankara. April 29, 2014.

19. If a Turkish citizen is a director of an excavation in Turkey, yet is working at an American institution (and living in the US and paying US taxes), the Turkish Ministry of Culture and Tourism still provides an excavation "dig house," guards, and nominal funds for annual excavation. In contrast, if a foreigner is an excavation director in Turkey, even if living in Turkey and working for a Turkish institution (and paying Turkish taxes), s/he receives none of these benefits.

20. Accessed February 10, 2018, https://eca.state.gov/cultural-heritage-center/ambassadors-fund-cultural-preservation.

21. Accessed February 10, 2018, https://tr.usembassy.gov/education-culture/grant-opportunities/afcp-small-grants-competition-2017.

BIBLIOGRAPHY

Acar, Özgen. "Canal Project Threatens Ancient Ephesus in Western Turkey." *Hurriyet Daily News*, October 23, 2016.

Adib, Khānum Hālidah. *The Turkish Ordeal*. New York: Century, 1928.

Akosy, Yaşar. *Love for 70 Years: Izmir Fair*. Izmir: Büyükşehir Belediyesi, 2001.

Aktürk, Şener. "The Fourth Style of Politics: Eurasianism as a Pro-Russian Rethinking of Turkey's Geopolitical Identity." *Turkish Studies* 16, no. 1 (2015): 54–79.

Aktürk, Şener. "Politics of History in Turkey: Revisionist Historiography's Challenge to the Official Version of the Turkish War of Liberation (1919–1922)." In *The Convolutions of Historical Politics,* edited by A. Miller and M. Lipman, 379–380. Budapest: Central European Press, 2012.

Allen, William Charles. *History of the United States Capitol: A Chronicle of Design, Construction, and Politics*. Vol. 106. Washington, DC: Government Printing Office, 2001.

American Anthropological Association. "Smithsonian Institution Foreign Currency Program in Archeology and Related Disciplines." *Anthropology News* 6, no. 8 (1965): 4–5.

Anderson, Benjamin. "'An Alternative Discourse': Local Interpreters of Antiquities in the Ottoman Empire." *Journal of Field Archaeology* 40, no. 4 (2015): 450–460.

Arredondo–Garrido, David. "References in Le Corbusier's Reorganization of Rural Habitat." *Journal of Architecture and Urbanism* 40, no. 2 (2016): 99–109.

Artukoglu, M. M., E. Tarkan, F. Gencler, and B. Miran. "Evaluating the Factors of Transition in Organic Cotton Production for Farmers: Case of Salihli, Turkey." *Bulgarian Journal of Agricultural Science* 15, no. 1 (2009): 77–83.

Atakuman, Çiğdem. "Value of Heritage in Turkey: History and Politics of Turkey's World Heritage Nominations." *Journal of Mediterranean Archaeology* 23, no. 1 (2010): 107–131.

Atakuman, Çiğdem. "Cradle or Crucible: Anatolia and Archaeology in the Early Years of the Turkish Republic (1923—1938)." *Journal of Social Archaeology* 8, no. 2 (2008): 214–235.

Atmiş, Erdoğan, Sezgin Özden, and Wietze Lise. "Public Participation in Forestry in Turkey." *Ecological Economics* 62, no. 2 (2007): 352–359.

Avcı, Duygu. "Mining Conflicts and Transformative Politics: A Comparison of Intag (Ecuador) and Mount Ida (Turkey) Environmental Struggles." *Geoforum* 84, August (2017): 316–325.

Aydin, Zülküf. "Neo-Liberal Transformation of Turkish Agriculture." *Journal of Agrarian Change* 10, no. 2 (2010): 149–187.

Baban, Feyzi, Suzan Ilcan, and Kim Rygiel. "Syrian Refugees in Turkey: Pathways to Precarity, Differential Inclusion, and Negotiated Citizenship Rights." *Journal of Ethnic and Migration Studies* 43, no. 1 (2017): 41–57.

Bacon, Mardges. "Le Corbusier and Postwar America." *Journal of the Society of Architectural Historians* 74, no. 1 (2015): 13–40.

Bacon, Mardges. *Le Corbusier in America: Travels in the Land of the Timid*. Cambridge, MA: MIT Press, 2001.

Bagchi, Amiya Kumar. *The Political Economy of Underdevelopment*. New York: Cambridge University Press, 1982.

Baird, Melissa F. "Heritage, Human Rights, and Social Justice." *Heritage and Society* 7, no. 2 (2014): 139–155.

Balkiz, Ozge. *Assessment of the Socio-Economic Values of Goods and Services Provided by Mediterranean Forest Ecosystems–Düzlerçami Forest, Turkey*. Valbonne: Plan Bleu, 2016.

Barr, James. *A Line in the Sand: Britain, France and the Struggle that Shaped the Middle East*. NewYork: Simon and Schuster, 2011.

Barrow, Christopher J. "River Basin Development Planning and Management: A Critical Review." *World Development* 26, no. 1 (1998): 171–186.

Barry, Andrew. *Material Politics: Disputes along the Pipeline*. West Sussex: John Wiley & Sons, 2013.

Basler, Teresa T., and David Wright. "The Making of a Collection: Mesoamerican Manuscripts at Princeton University." *Libraries and the Cultural Record* 43, no. 1 (2008): 29–55.

Beach, Vincent W. "Exploiting the Satellites." *Queen's Quarterly* 64, no. 1 (1957): 64.

Bennett, Tony. *The Birth of the Museum: History, Theory, Politics*. New York: Routledge, 2013.

Berber, Ferhat. "Kırımlı ve Kafkasyalı Muhacirlerin Manisa Kazasinda İskanı (1860–1876) [The Crimean and Caucasian Migrants in Manisa Province]." *Celal Bayar Üniversitesi Sosyal Bilimler Dergisi* 11, no. 3 (2013): 169–192.

Berberoglu, Berch. "State Capitalism and National Industrialization in Turkey." *Development and Change* 11, no. 1 (1980): 97–121.

Betts, Paul. "The Warden of World Heritage: UNESCO and the Rescue of the Nubian Monuments." *Past and Present* 226, no. suppl 10 (2015): 100–125.

Bevis, Teresa Brawner. *Higher Education Exchange between America and the Middle East through the Twentieth Century*. New York: Palgrave Macmillan, 2016.

Bilsel, Cana. "Ideology and Urbanism during the Early Republican Period: Two Master Plans for Izmir and Scenarios of Modernization." *Journal of the Faculty of Architecture* 16, no. 1–2 (1996): 13–60.

Birge, John Kingsley. *A Guide to Turkish Area Study*. Washington, D.C.: American Council of Learned Societies, Committee on Near Eastern Studies, 1949.

Bjørnlund, Matthias. "The 1914 Cleansing of Aegean Greeks as a Case of Violent Turkification." *Journal of Genocide Research* 10, no. 1 (2008): 41–58.

Bounni, Adnan. "Campaign and Exhibition from the Euphrates in Syria." *Annual of the American Schools of Oriental Research* 44 (1977): 1.

Bonini Baraldi, Sara, Daniel Shoup, and Luca Zan. "Understanding Cultural Heritage in Turkey: Institutional Context and Organisational Issues." *International Journal of Heritage Studies* 19, no. 7 (2013): 728–748.

Bochenski, Feliks, and William Diamond. "TVA's in the Middle East." *The Middle East Journal* vol. 4, no. 1 (1950): 52–82.

Bolca, Mustafa, Fulsen Özen, and Aslı Güneş. "Land use changes in Gediz Delta (Turkey) and their negative impacts on wetland habitats." *Journal of Coastal Research* 30, no. 4 (2012): 756–764.

Bouquet, Olivier. "Old Elites in a New Republic: The Reconversion of Ottoman Bureaucratic Families in Turkey (1909–1939)." *Comparative Studies of South Asia, Africa and the Middle East* 31, no. 3 (2011): 588–600.

Bozdoğan, Sibel. *Modernism and Nation Building: Turkish Architectural Culture in the Early Republic*. Seattle: University of Washington Press, 2001.

Bozdoğan, Sibel. "Architecture, Modernism and Nation–Building in Kemalist Turkey." *New Perspectives on Turkey* 10 (1994): 37–55.

Bozdoğan, Sibel, and Reşat Kasaba, eds. *Rethinking Modernity and National Identity in Turkey*. Vol. 7. Seattle: University of Washington Press, 1997.

Boztoprak, Tahsin, Osman Demir, and Yakup Emre Çoruhlu. "Comparison of Expropriation and Land Consolidation on the Regulation of Agricultural Land." *Sigma* 34, no. 1 (2016): 43–55.

Brew, John Otis. "Emergency Archaeology: Salvage in Advance of Technological Progress." *Proceedings of the American Philosophical Society* 105, no. 1 (1961): 1–10.

Brodie, Neil. "Syria and Its Regional Neighbors: A Case of Cultural Property Protection Policy Failure?" *International Journal of Cultural Property* 22, no. 2–3 (2015): 317–335.

Brown, L. Carl, ed. *Diplomacy in the Middle East: The International Relations of Regional and Outside Powers*. New York: IB Tauris, 2003.

Brown, Philip Marshall. "From Sevres to Lausanne." *American Journal of International Law* 18, no. 1 (1924): 113–116.

Burmeister, Gustave. "The Current and Prospective Foreign Trade Situation with Respect to Farm Products." *Journal of Farm Economics* 36, no. 5 (1954): 973–989.

Buckler, Sarah. "Imagined Communities Incorporated: Corporate Social Responsibility and Value Creation in a Globalised World." In *Corporate Social Responsibility: CSR, Sustainability, Ethics & Governance*, edited by S. Vertigans and S. Idowu, 3–22. Cham, Switzerland: Springer International Publishing, 2017.

Butler, Howard C. *Sardis I. The Excavations, Part 1: 1910–1914*. Leiden: E. J. Brill, Ltd, 1922.

Butler, Howard C. "Protection for the Historic Monuments and Objects of Art in Nearer Asia." *Art Bulletin* 2, no. 1 (1919): 46–48.

Canefe, Nergis. "The Politics of Public Protests against Extractivism in Turkey." *Peace Review* 28, no. 1 (2016): 10–19.

Celik, Zeynep. "Le Corbusier, Orientalism, Colonialism." *Assemblage* 17 (1992): 59–77.

Chapman, Michael E. "Taking Business to the Tiger's Gate: Thomas Handasyd Perkins and the Boston-Smyrna-Canton Opium Trade of the Early Republic." *Journal of the Royal Asiatic Society Hong Kong Branch* 52 (2012): 7–28.

Clarke, John, Dave Bainton, Noémi Lendvai, and Paul Stubbs. *Making Policy Move: Towards a Politics of Translation and Assemblage*. Bristol: Policy Press, 2015.

Clavin, Patricia. *Securing the World Economy: The Reinvention of the League of Nations, 1920–1946*. New York: Oxford University Press, 2013.

Clavin, Patricia, and Jens-Wilhelm Wessel. "Transnationalism and the League of Nations: Understanding the Work of Its Economic and Financial Organisation." *Contemporary European History* 14, no. 4 (2005): 465–492.

Cochrane, Willard W. "Public Law 480 and Related Programs." *The ANNALS of the American Academy of Political and Social Science* 331, no. 1 (1960): 14–19.

Coombe, Rosemary J., and Melissa F. Baird. "The Limits of Heritage: Corporate Interests and Cultural Rights on Resource Frontiers." In *A Companion to Heritage Studies,* 1st ed., edited by W. Logan, M. N. Craith, and W. Kockel, 337–354. West Sussex: Wiley & Sons Publishing, 2016.

Coombe, Rosemary J., and Lindsay M. Weiss. "Neoliberalism, Heritage Regimes, and Cultural Rights." In *Global Heritage: A Reader*, edited by L. Meskell, 43–69. West Sussex: Wiley-Blackwell, 2015.

Coş, Kıvanç, and Pinar Bilgin. "Stalin's Demands: Constructions of the 'Soviet Other' in Turkey's Foreign Policy, 1919–1945." *Foreign Policy Analysis* 6, no. 1 (2010): 43–60.

Criss, Nur Bilge. "Strategic Nuclear Missiles in Turkey: The Jupiter Affair, 1959–1963." *Journal of Strategic Studies* 20, no. 3 (1997): 97–122.

Crozier, Andrew J. "The Establishment of the Mandates System 1919–25: Some Problems Created by the Paris Peace Conference." *Journal of Contemporary History* 14, no. 3 (1979): 483–513.

Culvahouse, Tim, ed. *The Tennessee Valley Authority: Design and Persuasion*. Vol. 8. New York: Princeton Architectural Press, 2007.

Çelik, Zeynep. *Displaying the Orient: Architecture of Islam at Nineteenth-Century World's Fairs*. Vol. 12. Los Angeles: University of California Press, 1992.

Çiftci, Gülsel, Gülay Çakır, and Ali Çakır. "Political Crisis' and Their Implications on Heritage Sites: The Case of Turkey." *Journal of Tourism Theory and Research* 2, no. 2 (2016): 126–138.

Davies, Thomas. *NGOs: A New History of Transnational Civil Society*. New York: Oxford University Press, 2014.

Davis, Jack L. "The American School of Classical Studies and the Politics of Volunteerism." *Hesperia: The Journal of the American School of Classical Studies at Athens* 82, no. 1 (2013): 15–48.

Davis, Jack L. "A Foreign School of Archaeology and the Politics of Archaeological Practice: Anatolia, 1922." *Journal of Mediterranean Archaeology* 16, no. 2 (2003): 145–172.

Davis, Jack L., and Natalia Vogeikoff-Brogan, eds. *Philhellenism, Philanthropy, or Political Convenience? American Archaeology in Greece*. Athens: American School of Classical Studies, 2013.

Davis, Jack L., and Natalia Vogeikoff-Brogan. "Introduction: Philhellenism, Philanthropy, or Political Convenience?" *Hesperia: The Journal of the American School of Classical Studies at Athens* 82, no. 1 (2013): 1–14.

Deans, Robert H. "US Foreign Assistance Programs: The Impact of Local Currencies in Economic Development." *Journal of Developing Areas* 5, no. 4 (1971): 589–604.

Demir, Osman, and Yakup Emre Çoruhlu. "Determining the Property Ownership on Cadastral Works in Turkey." *Land Use Policy* 26, no. 1 (2009): 112–120.

Demir, Osman, Bayram Uzun, and Mehmet Çete. "Turkish Cadastral System." *Survey Review* 40, no. 307 (2008): 54–66.

Demirdöğen, Alper, Emine Olhan, and Jean-Paul Chavas. "Food vs. Fiber: An Analysis of Agricultural Support Policy in Turkey." *Food Policy* 61 (2016): 1–8.

Demirli, Meltem Eranıl, Zeynep Tuna Ultav, and Neslihan Demirtaş-Milz. "A Socio–Spatial Analysis of Urban Transformation at a Neighborhood Scale: The Case of the Relocation of Kadifekale Inhabitants to TOKİ Uzundere in Izmir." *Cities* 48 (2015): 140–159.

Demiryürek, K., C. Stopes, and A. Güzel, "Organic Agriculture: The Case of Turkey." *Outlook on Agricultrue* 37, no. 4 (2008): 261–267.

DeNovo, John A. *American Interests and Policies in the Middle East, 1900–1939*. Minneapolis: University of Minnesota Press, 1963.

Detweiler, A. H. "Some Early Jewish Architectural Vestiges from Jerash." *Bulletin of the American Schools of Oriental Research* 87 (1942): 10–17.

Detweiler, A. Henry. *Manual of Archaeological Surveying*. Vol. 2. New Haven, CT: American Schools of Oriental Research, 1948.

Dockrill, Michael L. "Historical Note: The Foreign Office and the Proposed Institute of International Affairs 1919." *International Affairs (Royal Institute of International Affairs 1944–)* 56, no. 4 (1980): 665–672.

Dolan, Catherine, and Dinah Rajak, eds. *The Anthropology of Corporate Social Responsibility*. Vol. 18. New York: Berghahn Books, 2016.

Domhardt, Konstanze Sylva. "The Garden City Idea in the CIAM Discourse on Urbanism: A Path to Comprehensive Planning." *Planning Perspectives* 27, no. 2 (2012): 173–197.

Dornemann, Rudolph H. "Tell Hadidi: One Bronze Age Site among Many in the Tabqa Dam Salvage Area." *Bulletin of the American Schools of Oriental Research* (1988): 13–42.

Dornemann, Rudolph H. "Salvage Excavations at Tell Hadidi in the Euphrates River Valley." *The Biblical Archaeologist*, 48, no. 1 (1985): 49–59.

Douet, James, ed. *Industrial Heritage Re-Tooled: The TICCIH Guide to Industrial Heritage Conservation.* Walnut Creek, CA: Left Coast Press, 2013.

Dougherty, James E. "The Aswan Decision in Perspective." *Political Science Quarterly* 74, no. 1 (1959): 21–45.

Dougherty, Michael L., Sarah Anderson, and Manuel Perez-Rocha, "The Rise of the Corporate Investment Rights Regime and 'Extractive Exceptionalism': Evidence from El Salvador." In *Mining in Latin America: Critical Approaches to the New Extraction*, edited by. K. Deonandan and M. L. Dougherty, 226–249. London: Routledge, 2016.

Downs, Jacques M. "American Merchants and the China Opium Trade, 1800–1840." *Business History Review* 42, no. 4 (1968): 418–442.

Dülger, Murat Volkan. "The Balance between the Public Interest and the Protection of Property Rights: A Judgment by the European Court of Human Rights About Expropriation." February 5, 2015 (accessed January 2017). https://ssrn.com/abstract=2560894.

Durgun, Sezgi. "Cultural Cold War at the Izmir International Fair: 1950s–60s." In *Turkey in the Cold War*, edited by C. Örnek and Ç. Üngör, 67–86. New York: Palgrave Macmillan UK, 2013.

Dyson, Stephen L. *Ancient Marbles to American Shores: Classical Archaeology in the United States*. Philadelphia: University of Pennsylvania Press, 1998.

Eckes Alfred E., Jr. *The United States and the Global Struggle for Minerals*. Austin: University of Texas Press, 1979.

Ekbladh, David. "Meeting the Challenge from Totalitarianism: The Tennessee Valley Authority as a Global Model for Liberal Development, 1933–1945." *The International History Review* 32, no. 1 (2010): 47–67.

Ekbladh, David. "'Mr. TVA': Grass-Roots Development, David Lilienthal, and the Rise and Fall of the Tennessee Valley Authority as a Symbol for US Overseas Development, 1933–1973." *Diplomatic History* 26, no. 3 (2002): 335–374.

Eldem, Edhem. "From Blissful Indifference to Anguished Concern: Ottoman Perceptions of Antiquities, 1799–1869." In *Scramble for the Past: A Story of Archaeology in the Ottoman Empire, 1914*, edited by Z. Bahrani, Z. Çelik, and E. Eldem, 281–330. Istanbul: SALT, 2011.

Eldem, Edhem. "The Undesirables of Smyrna, 1926." *Mediterranean Historical Review* 24, no. 2 (2009): 223–227.

Eldem, Edhem, Daniel Goffman, and Bruce Masters. *The Ottoman City between East and West: Aleppo, Izmir, and Istanbul*. New York: Cambridge University Press, 1999.

Emecen, Feridun Mustafa. *XVI. Asırda Manisa Kazası* [Manisa Province in the 16th Century]. Türk Tarih Kurumu Basımevi, Istanbul 1989.

Emek, Uğur. "Turkish Experience with Public Private Partnerships in Infrastructure: Opportunities and Challenges." *Utilities Policy* 37 (2015): 120–129.

Emek, Ugur, and Muhittin Acar. "6. Public Procurement in Infrastructure: The Case of Turkey." *Government Favouritism in Europe* (2015): 84.

Einstein, Albert, Sigmund Freud, Fritz Moellenhoff, and Anna Moellenhoff. *Why War? The Correspondence between Albert Einstein and Sigmund Freud*. Chicago Institute for Psychoanalysis, 1933.

Erdönmez, Cihan, and Aynur Aydin Coşkun. "Sustainability in Turkish Forest Legislation and Administration." Accessed April 2017. http://www.fao.org/docrep/ARTICLE/WFC/XII/0072-C1.HTM.

Ergül, Teoman. *Kurtuluş Savaşında Manisa, 1919–1922* [Manisa during the Turkish War of Independence, 1919–1922]. Vol. 1. Manisa: Manisa Kültür Sanat Kurumu, 1991.

Essig, Linda. "Means and Ends: A Theory Framework for Understanding Entrepreneurship in the US Arts and Culture Sector." *The Journal of Arts Management, Law, and Society* 45, no. 4 (2015): 227–246.

Essig, Linda. "Same or Different? The 'Cultural Entrepreneurship' and 'Arts Entrepreneurship' Constructs in European and US Higher Education." *Cultural Trends* (2017): 1–13.

Ferguson, James. *The Anti-Politics Machine: 'Development,' Depoliticization and Bureaucratic Power in Lesotho*. New York: Cambridge University Press, 1990.

Ferry, Elizabeth. "Gold Prices as Material-Social Actors: The Case of the London Gold Fix." *The Extractive Industries and Society* 3, no. 1 (2016): 82–85.

Finefrock, Michael M. "Laissez-Faire, the 1923 Izmir Economic Congress and Early Turkish Developmental Policy in Political Perspective." *Middle Eastern Studies* 17, no. 3 (1981): 375–392.

Finefrock, Michael M. "Atatürk, Lloyd George and the Megali Idea: Cause and Consequence of the Greek Plan to Seize Constantinople from the Allies, June–August 1922." *Journal of Modern History* 52, no. S1 (1980): D1047–D1066.

Fisher, Christopher T. "'Moral Purpose Is the Important Thing': David Lilienthal, Iran, and the Meaning of Development in the US, 1956–63." *International History Review* 33, no. 3 (2011): 431–451.

Fındıkoğlu, Ziyaeddin Fahri. "Rus Kazakları Etrafında Nazariyeler ve Bir Ek [Theories about the Russian Cossacks and an Appendix]." *Iktisat Fakültesi Mecmuası* 23, no. 3–4 (1963): 151–166.

Flam, Harry. "Turkey and the EU: Politics and Economics of Accession." *CESifo Economic Studies* 50, no. 1 (2004): 171–210.

Floor, Willem. "From Anatolia to Indonesia: Opium Trade and the Dutch Community of Izmir, 1820–1940." *Journal of the American Oriental Society* 120, no. 4 (2000): 690.

Forsythe, James L. "World Cotton Technology since World War II." *Agricultural History* 54, no. 1 (1980): 208–222.

Foss, Clive. *Byzantine and Turkish Sardis*. No. 4. Cambridge, MA: Harvard University Press, 1976.

Frangakis-Syrett, Elena. "Capital Accumulation and Family Business Networks in Late Ottoman Izmir." *International Journal of Turkish Studies* 21, no. 1/2 (2015): 55–72.

Frangakis-Syrett, Elena. "British economic activities in Izmir in the second half of the nineteenth and early twentieth centuries." *New Perspectives on Turkey* 6, Fall (1991): 191–227.

Frangakis-Syrett, Elena. "Banking in Izmir in the Early Twentieth Century." *Mediterranean Historical Review* 24, no. 2 (2009): 115–132.

Frangakis-Syrett, Elena. "The Making of an Ottoman Port: The Quay of Izmir in the Nineteenth Century." *The Journal of Transport History* 22, no. 1 (2001): 23–46.

Frangakis-Syrett, Elena. "The Economic Activities of the Greek Community of Izmir in the Second Half of the Nineteenth and Early Twentieth Centuries." In *Ottoman Greeks in the Age of Nationalism: Politics, Economy, and Society in the Nineteenth Century*, edited by D. Gondicas and C. Issawi, 17–44. Princeton, NJ: Darwin Press, 1999.

Frangakis-Syrett, Elena. "American Trading Practices in Izmir in the Late Nineteenth and Early Twentieth Centuries." *Histoire économique et sociale de l'Empire ottoman et de la Turquie (1326–1960): actes du sixième congrès international tenu à Aix-en-Provence du 1er au 4 juillet 1992* 8 (1995): 177.

Frangakis-Syrett, Elena. "Implementation of the 1838 Anglo-Turkish Convention on Izmir's Trade: European and Minority Merchants." *New Perspectives on Turkey* 7 (1992): 91–112.

Frangakis-Syrett, Elena. "British Economic Activities in Izmir in the Second Half of the Nineteenth and Early Twentieth Centuries." *New Perspectives on Turkey* 6 (1991): 191–227.

Frangakis-Syrett, Elena. "Trade between Ottoman Empire and Western Europe: The Case of Izmir in the Eighteenth Century." *New Perspectives on Turkey* 2 (1988): 1–18.

Frothingham, Donald. "New Trade Route Open to Turkey: Big Help to US." *Chicago Daily Tribune*, April 6, 1914, p. 13.

Gardiner, Arthur Z. "Point Four and the Arab World: An American View." *The Middle East Journal* (1950): 296–306.

Gekas, Athanasios. "Class and Cosmopolitanism: The Historiographical Fortunes of Merchants in Eastern Mediterranean Ports." *Mediterranean Historical Review* 24, no. 2 (2009): 95–114.

Gill, Rebecca. "'Now I have seen evil, and I cannot be silent about it': Arnold J. Toynbee and His Encounters with Atrocity, 1915–1923." In *Evil, Barbarism and Empire: Britain and Abroad, C.1830–2000*, edited by T. Crook, R. Gill, and B. Taithe,172–200. London: Palgrave Macmillan UK, 2011.

Gissen, David. "The Architectural Production of Nature, Dendur/New York." *Grey Room* 34 (2009): 58–79.

Gisselquist, David, and Carl Pray. "The Impact of Turkey's 1980s Seed Regulatory Reform." *Easing Barriers to Movement of Plant Varieties for Agricultural Development, Págs* (1997): 113–131.

Giuliani, Elisa. "Human Rights and Corporate Social Responsibility in Developing Countries' Industrial Clusters." *Journal of Business Ethics* 133, no. 1 (2016): 39–54.

Glendinning, Miles. *The Conservation Movement: A History of Architectural Preservation: Antiquity to Modernity*. New York: Routledge, 2013, 198–201.

Glueck, Nelson. "Jerash in the Spring of 1933. Preliminary Report of the Joint Expedition of the School in Jerusalem and Yale University." *Bulletin of the American Schools of Oriental Research* 53 (1934): 2–13.

Godard, Louis. "L'irrigation en Turquie [Irrigation in Turkey]" *Le Génie Civil* 56 (1909–10): 283.

Goffman, Daniel. "Jews in Early Modern Ottoman Commerce." *Jews, Turks and Ottomans* (2002): 15–34.

Goffman, Daniel. *Izmir and the Levantine World: 1550–1650* (Vol. 5). Seattle: University of Washington Press, 1990.

Gökmen, Ertan. "1888 Tarihli Akhisar Zımpara Madeni Imtiyaz [1888 Concession of the Zımpara Mine, Akhisar]." *Sosyal Bilimler* 2, no. 1 (2004): 53–64.

Gold, John R. *The Experience of Modernism: Modern Architects and the Future City, 1928–53*. New York: Taylor & Francis, 2013.

Goldstein, Erik. "The Foreign Office and Political Intelligence 1918–1920." *Review of International Studies* 14, no. 04 (1988): 275–288.

Goldstein, Erik. "British Peace Aims and the Eastern Question: The Political Intelligence Department and the Eastern Committee, 1918." *Middle Eastern Studies* 23, no. 4 (1987): 419–436.

Goode, James F. *Negotiating for the Past: Archaeology, Nationalism, and Diplomacy in the Middle East, 1919—1941*. Austin: University of Texas Press, 2007.

Grabill, Joseph L. "Missionary Influence on American Relations with the Near East, 1914–1923." *The Muslim World* 58, no. 2 (1968): 141–154.

Grabill, Joseph L. "Cleveland H. Dodge, Woodrow Wilson, and the Near East." *Journal of Presbyterian History (1962–1985)* 48, no. 4 (1970): 249–264.

Grathwol, Robert P., and Donita M. Moorhus. *Bricks, Sand, and Marble: US Army Corps of Engineers Construction in the Mediterranean and Middle East, 1947–1991*. Vol. 45. Center of Military History, Corps of Engineers, US Army, 2009.

Greenewalt, Crawford Hallock, ed. *The City of Sardis: Approaches in Graphic Recording*. Cambridge, MA: Harvard University Art Museum, 2003.

Gur, Faik. "Sculpting the Nation in Early Republican Turkey." *Historical Research* 86 no. 232 (2013): 342–372.

Gül, Murat. *Emergence of Modern Istanbul: Transformation and Modernisation of a City*. New York: IB Tauris, 2009.

Gülmez, Nilay Ünsal. "Ege Taşrasında 'Modern' in Mitleşmesi: Kemer (Bozdoğan, Aydın) Ve Demirköprü (Salihli, Manisa) Baraj Siteleri Ve Lojmanları [The Mythologisation of the "The Modern" in the Rural Aegean: The Case of Kemer (Bozdoğan, Aydın, and Demirköprü (Salihli, Manisa) Dam Sites and Lodgings]." *METU Journal of the Faculty of Architecture* 32, no. 2 (2016): 19–43.

Gümüş, Cantürk. "Historical Development of Forestry Education in the Context of Forest Resources Management in Turkey." *Turkish Journal of Forestry* 17, no. 1 (2016): 93–98.

Gürel, Perin E. *Limits of Westernization: A Cultural History of America in Turkey*. New York: Columbia University Press, 2017.

Habib, Salem Nasr. "Methods of Capturing the Potential Benefits of the Aswan High Dam in Egypt, UAR." PhD diss., Georgia Institute of Technology, 1961.

Hanfmann, George Maxim Anossov, and Ilse Hanfmann. *Letters from Sardis*. Cambridge, MA: Harvard University Press, 1972.

Hanfmann, George. *Sardis from Prehistoric to Roman Times: Results of the Archaeological Exploration of Sardis, 1958–1975*. Cambridge, MA: Harvard University Press, 1983.

Hale, William. "Ideology and Economic Development in Turkey 1930–1945." *British Journal of Middle Eastern Studies* 7, no. 2 (1980): 100–117.

Harmancioglu, Nilgun B., Kurt Fedra, and Filiz Barbaros. "Analysis for Sustainability in Management of Water Scarce Basins: The Case of the Gediz River Basin in Turkey." *Desalination* 226, no. 1–3 (2008): 175–182.

Harris, Leila M., and Samer Alatout. "Negotiating Hydro-Scales, Forging States: Comparison of the Upper Tigris/Euphrates and Jordan River Basins." *Political Geography* 29, no. 3 (2010): 148–156.

Harrison, Rodney. "Beyond 'Natural' and 'Cultural' Heritage: Toward an Ontological Politics of Heritage in the Age of Anthropocene." *Heritage & Society* 8, no. 1 (2015): 24–42.

Hasluck, F. W. "Unpublished inscriptions from Cyzicus Neighbourhood." *Journal of Hellenic Studies* 24 (1904): 20–40.

Hassan, Fekri A. "The Aswan High Dam and the International Rescue Nubia Campaign." *African Archaeological Review* 24, no. 3–4 (2007): 73–94.

Hasselskog, Malin. "Rwandan Developmental 'Social Engineering': What Does It Imply and How Is It Displayed?" *Progress in Development Studies* 15, no. 2 (2015): 154–169.

Hasselskog, Malin, Peter J. Mugume, Eric Ndushabandi, and Isabell Schierenbeck. "National Ownership and Donor Involvement: An Aid Paradox Illustrated by the Case of Rwanda." *Third World Quarterly* (2016): 1–15.

Hastaoglou-Martinidis, Vilma. "Urban Aesthetics and National Identity: The Refashioning of Eastern Mediterranean Cities between 1900 and 1940." *Planning Perspectives* 26, no. 2 (2011): 153–182.

Hershlag, Zvi Yehuda. *Introduction to the Modern Economic History of the Middle East*. Leiden: Brill, 1980.

Herzfeld, Michael. *Evicted from Eternity: The Restructuring of Modern Rome*. Chicago: University of Chicago Press, 2009.

Hirst, Samuel J. "Anti-Westernism on the European Periphery: The Meaning of Soviet-Turkish Convergence in the 1930s." *Slavic Review* 72, no. 1 (2013): 32–53.

Hobsbawm, Eric, and Terence Ranger, eds. *The Invention of Tradition*. New York: Cambridge University Press, [1983] 2012.

Horvath, Janos. "Economic Aid Flow from the USSR: A Recount of the First Fifteen Years." *Slavic Review* 29, no. 4 (1970): 613–632.

Human, Helen. "Ethical or Empty Gestures? World Heritage Nominations in Turkey." In *Heritage in Action*, edited by H. Silverman, E. Waterton, and S. Watson, 75–89. Cham, Switzerland: Springer International Publishing, 2017.

Human, Helen. "Democratising World Heritage: The Policies and Practices of Community Involvement in Turkey." *Journal of Social Archaeology* 15, no. 2 (2015): 160–183.

Humphrey, Llewellyn Mealey. *Mexican Wheat Comes to Turkey*. United States Agency for International Development, Food and Agricultural Division, 1969.

Huxley, Julian. *TVA, Adventure in Planning*. Kingston on Thames: Knapp, Drewett & Sons Ltd., 1943.

Içduygu, Ahmet, and Doğuş Şimşek. "Syrian Refugees in Turkey: Towards Integration Policies." *Turkish Policy Quarterly* 15, no. 3 (2016): 59–69.

Ikeotuonye, Festus C.R.A. "Lateral Shades of Social Engineering: A Critical Exploration of 'Interest Representation', 'State' and 'Development'." *Community Development Journal* 37, no. 1 (2002): 69–79.

International Bank for Reconstruction and Development. Eastern European Papers, No. 5. Preliminary Report on Turkey, Part 1 (March 1948).

Ivusic, Christopher W. "Training Experts of Other Nations." *Reclamation ERA* 54, no. 4 (1968): 92.

Johnson, Kendall. "Henry James and the China Trade." *MFS Modern Fiction Studies* 60, no. 4 (2014): 677–710.

Kacar, Duygu. "A Unique Spatial Practice for Transforming the Social and Cultural Patterns: Atatürk Forest Farm in Ankara (1)." *METU JFA* 1 (2011): 165–178.

Kalkan, Mustafa. "Sovyetler Döneminde Kazakların Göç Hareketleri ve Anadolu'da (Altay Köyü'nde) İskan Edilişeli [Emigration Movements of the Kazakh Turks in the Soviet Era and their Settlement in Anatolia (the Village of Altay)]." *TÜBAR*–XXI (2007): 123–141.

Kalyukin, Alexander, Thomas Borén, and Andrew Byerley. "The Second Generation of Post-Socialist Change: Gorky Park and Public Space in Moscow." *Urban Geography* 36, no. 5 (2015): 674–695.

Karson, Lawrence. *American Smuggling as White Collar Crime*. New York: Routledge, 2014.

Karaosmanoğlu, Attila. *Izmir Karşıyaka'dan dünya'ya* [From Izmir's Karşıyaka to the World]. Vol. 890. Türkiye İş Bankası Kültür Yayınları, 2005.

Kasaba, Reşat "Izmir 1922: A Port City Unravels." In *Modernity & Culture: From the Mediterranean to the Indian Ocean*, edited by L. T. Fawaz and C. A. Bayly, 204–229. New York: Columbia University Press, 2002.

Kayaoğlu, Barın. "Strategic Imperatives, Democratic Rhetoric: The United States and Turkey, 1945–52: Cold War in the Aegean." *Cold War History* 9, no. 3 (2009): 321–345.

Kenanoğlu, Zerrin, and Özlem Karahan. "Policy Implementations for Organic Agriculture in Turkey." *British Food Journal* 104, no. 3/4/5 (2002): 300–318.

Kersel, Morag M., Christina Luke, and Christopher H. Roosevelt. "Valuing the Past Perceptions of Archaeological Practice in Lydia and the Levant." *Journal of Social Archaeology* 8, no. 3 (2008): 298–319.

Keskinok, H. Çağatay. "Urban Planning Experience of Turkey in the 1930s." *METU JFA* 27, no. 2 (2010): 173–188.

Kirchmann, Holger, Gudni Thorvaldsson, Lars Bergström, Martin Gerzabek, Olof Andrén, Lars-Olov Eriksson, and Mikael Winninge. "Fundamentals of organic agriculture–Past and present." In *Organic Crop Production–Ambitions and Limitations*, edited by H. Kirchmann and L. Bergstrom, 13–37. Netherlands: Springer, 2008.

Knappen, Tippetts, Abbett Engineering Company of New York. *Gediz Basin Study*. 1950.

Knox, Sanka. "Archaeologists Find Huge Tomb of Lydian King." *New York Times*, November 8, 1964.

Kolluoğlu, Biray "Cityscapes and Modernity: Smyrna Morphing into Izmir." In *Ways to Modernity in Greece and Turkey: Encounters with Europe, 1850–1950*," edited by A. Frangoudaki and C. Keyder, 217–235. London: IB Tauris Publishers, 2007.

Koknar, Ali M. "Turkey's security relations in Central Asia." In *Eurasia in Balance: The US and the Regional Power Shift*, edited by A. Cohen, 101–126. New York: Routledge, 2017.

Kolars, John F., and William A. Mitchell. *The Euphrates River and the Southeast Anatolia Development Project*. Carbondale: Southern Illinois University Press, 1991.

Kolko, Gabriel. *The Triumph of Conservatism: A Reinterpretation of American History, 1900–1918*. New York: Free Press, 1963.

Kolluoğlu-Kırlı, Biray. "Forgetting the Smyrna Fire." *History Workshop Journal* 60, no. 1 (2005): 25–44.

Kolluoğlu-Kırlı, Biray. "The Play of Memory, Counter-Memory: Building Izmir on Smyrna's Ashes." *New Perspectives on Turkey* 26 (2002): 1–28.

Konak, Nahide. "Ecological Modernization and Eco-Marxist Perspectives: Globalization and Gold Mining Development in Turkey." *Capitalism Nature Socialism* 19, no. 4 (2008): 107–130.

Kortan, Enis. *Turkish Architecture and Urbanism through the Eyes of Le Corbusier*. Ankara: Middle East Technical University, 1991.

Kozat, Burcak Keskin. *Negotiating Modernization through US Foreign Assistance: Turkey's Marshall Plan (1948–1952) Re-interpreted*. PhD dissertation (Sociology), University of Michigan, 2006.

Kraeling, Carl Hermann. *Gerasa, City of the Decapolis: An Account Embodying the Record of a Joint Excavation Conducted by Yale University and the British School of Archaeology in Jerusalem (1928–1930), and Yale University and the American Schools of Oriental Research (1930–1931, 1933–1934)*. New Haven, CT: American Schools of Oriental Research, 1938.

Lafrenz Samuels, Kathryn. "Transnational Turns for Archaeological Heritage: From Conservation to Development, Governments to Governance." *Journal of Field Archaeology* 41, no. 3 (2016): 355–367.

Lagendijk, Vincent. "Divided Development: Post-War Ideas on River Utilisation and Their Influence on the Development of the Danube." *International History Review* 37, no. 1 (2015): 80–98.

Laine, Jussi P., and Anna Casaglia. "Challenging Borders: A Critical Perspective on the Relationship between State, Territory, Citizenship and Identity: Introduction." *Europa Regional* 24, no. 1/2 (2017): 3–8.

Lalaki, Despina. "On the Social Construction of Hellenism Cold War Narratives of Modernity, Development and Democracy for Greece." *Journal of Historical Sociology* 25, no. 4 (2012): 552–577.

Lane, Paul J. "Archaeology in the Age of the Anthropocene: A Critical Assessment of Its Scope and Societal Contributions." *Journal of Field Archaeology* 40, no. 5 (2015): 485–498.

Larson, Donald, Will Martin, Sebnem Sahin, and Marinos Tsigas. "Agricultural Policies and Trade Paths in Turkey." *World Economy* (2016): 1194–1224.

Larson, Victoria Tietze. "Classics and the Acquisition and Validation of Power in Britain's "Imperial Century" (1815–1914)." *International Journal of the Classical Tradition* 6, no. 2 (1999): 185–225.

Lauren, Paul Gordon. "Human Rights in History Diplomacy and Racial Equality at the Paris Peace Conference." *Diplomatic History* 2, no. 3 (1978): 257–278.

Lendvai, Noemi, and Paul Stubbs. "Assemblages, Translation, and Intermediaries in South East Europe: Rethinking Transnationalism and Social Policy." *European Societies* 11, no. 5 (2009): 673–695.

Liakos, Antonis. "Historical Time and National Space in Modern Greece." In *Regions in Central and Eastern Europe: Past and Present*, edited by T. Hayashi and H. Fukuda, 205–227. Hokkaido: Hokkaido University, 2007.

Liapis, Vayos. "'The Painful Memory of Woe': Greek Tragedy and the Greek Civil War in the Work of George Seferis." *Classical Receptions Journal* 6, no. 1(2014): 74–103.

Lilienthal, David E. *Democracy on the March*. New York: Harper & Brothers, 1944.

Little, Douglas. "Cold War and Covert Action: The United States and Syria, 1945–1958." *Middle East Journal* 44, no. 1 (1990): 51–75.

Little, Douglas. "The New Frontier on the Nile: JFK, Nasser, and Arab Nationalism." *Journal of American History* 75, no. 2 (1988): 501–527.

Livingston, Craig. "'One Thousand Wings': The United States Air Force Group and the American Mission for Aid to Turkey, 1947–50." *Middle Eastern Studies* 30, no. 4 (1994): 778–825.

Luke, Christina. "The 40th World Heritage Session in Istanbul, Turkey: A Reflection on the Legacies of Heritage Policy and Missed Mega-Heritage." *Journal of Field Archaeology* 41 No. 6 (2016): 641–644.

Luke, Christina. "Cultural Sovereignty in the Balkans and Turkey: The Politics of Preservation and Rehabilitation." *Journal of Social Archaeology* 13, no. 3 (2013): 350–370.

Luke, Christina. "Archaeological Heritage and the Turkish War of Independence in Central Lydia." *Museum International* 62, no. 1-2 (2010): 42–47.

Luke, Christina, and Elvan Cobb. "Dwelling in Hacıveliler: Social-Engineering Policies in the Context of Space, Place and Landscape in Rural, Western Turkey." *Anatolian Studies* 63 (2013): 155–173.

Luke, Christina, and Morag Kersel. *US Cultural Diplomacy and Archaeology: Soft Power, Hard Heritage*. New York: Routledge, 2013.

Luke, Christina, and Christopher H. Roosevelt. "Cup-Marks and Citadels: Evidence for Libation in the Second-Millennium BCE Marmara Lake Basin, Western Anatolia." *Bulletin of the American Schools of Oriental Research* 378 (2017): 1–23.

Luke, Christina, and Christopher H. Roosevelt. "Memory and Meaning in Bin Tepe, the Lydian Cemetery of the 'Thousand Mounds'." *Tumulus as Sema: Space, Politics, Culture and Religion in the First Millennium BC* 27 (2016): 407–428.

Luke, Christina, Christopher H. Roosevelt, and Catherine B. Scott. "Yörük Legacies: Space, Scent, and Sediment Geochemistry." *International Journal of Historical Archaeology* (2016): 1–26.

Kuş, Başak. "Weak States, Unruly Capitalists, and the Rise of Étatism in Late Developers: The Case of Turkey." *British Journal of Middle Eastern Studies* 42, no. 3 (2015): 358–374.

Macekura, Stephen. "The Point Four Program and US International Development Policy." *Political Science Quarterly* 128, no. 1 (2013): 127–160.

Macfie, A. L. "The Revision of the Treaty of Sèvres: The First Phase (August 1920–September 1922)." *Balkan Studies* 24 (1983): 57–88.

Manela, Erez. *The Wilsonian Moment: Self-Determination and the International Origins of Anticolonial Nationalism*. Oxford University Press on Demand, 2007.

Mansel, Philip. "Cities of the Levant—The Past for the Future?" *Asian Affairs* 45, no. 2 (2014): 220–242.

McEwan, Cheryl, Emma Mawdsley, Glenn Banks, and Regina Scheyvens. "Enrolling the Private Sector in Community Development: Magic Bullet or Sleight of Hand?" *Development and Change* 48, no. 1 (2017): 28–53.

Meskell, Lynn. *A Future in Ruins: UNESCO, World Heritage, and the Dream of Peace*. London: Oxford University Press, 2018.

Meskell, Lynn. "Gridlock: UNESCO, Global Conflict and Failed Ambitions." *World Archaeology* 47, no. 2 (2015): 225–238.

Meskell, Lynn. "Transacting UNESCO World Heritage: Gifts and Exchanges on a Global Stage." *Social Anthropology* 23, no. 1 (2015): 3–21.

Meskell, Lynn. "States of Conservation: Protection, Politics, and Pacting within UNESCO's World Heritage Committee." *Anthropological Quarterly* 87, no. 1 (2014): 217–243.

Meskell, Lynn. "From Paris to Pontdrift: UNESCO Meetings, Mapungubwe and Mining." *The South African Archaeological Bulletin* (2011): 149–156.

Meskell, Lynn. "Human Rights and Heritage Ethics." *Anthropological Quarterly* 83, no. 4 (2010): 839–859.

Meskell, Lynn. "The Practice and Politics of Archaeology in Egypt." *Annals of the New York Academy of Sciences* 925, no. 1 (2000): 146–169.

Minoglou, Ioanna Pepelasis. "Ethnic Minority Groups in International Banking: Greek Diaspora Bankers of Constantinople and Ottoman State Finances, c. 1840–81." *Financial History Review* 9, no. 2 (2002): 125–146.

Mitchell, Timothy. *Rule of Experts: Egypt, Techno-Politics, Modernity.* Los Angeles: University of California Press, 2002.

Mitten, David G., and G. Yüğrüm. "Excavations at Ahlatlı Tepecik on the Gygean Lake, 1968." *Türk Arkeoloji Dergisi* 17 (1968): 125–131.

Mitten, David G., and G. Yüğrüm. "The Gygean Lake, 1969: Eski Balıkhane, Preliminary Report." *Harvard Studies in Classical Philology* 75 (1971): 191–195.

Mitten, David G., and G. Yüğrüm. "Ahlatlı Tepecik beside the Gygean Lake," *Archaeology* 27 (1974): 22–29.

Montgomery, A. E. "The Making of the Treaty of Sèvres of 10 August 1920." *The Historical Journal* 15, no. 4 (1972): 775–787.

Moore, Andrew M. T., G. C. Hillman, and A. J. Legge. "The Excavation of Tell Abu Hureyra in Syria: A Preliminary Report." *Proceedings of the Prehistoric Society*, vol. 41, December (1975): 50–77.

Morack, Ellinor. "Fear and Loathing in 'Gavur' Izmir: Emotions in Early Republication Memories of the Greek Occupation (1919–22)." *International Journal of Middle East Studies* 49, no. 1 (2017): 71–89.

Mumford, Eric Paul. *The CIAM Discourse on Urbanism, 1928–1960.* Cambridge, MA: MIT Press, 2002.

Mumford, Eric. "CIAM Urbanism after the Athens Charter." *Planning Perspective* 7, no. 4 (1992): 391–417.

Munro, J. A. R. "Dascylium." *Journal of Hellenic Studies* 32 (1912): 57–67.

McKeown-Ice, Rosalyn. "Rostov Region, Russia-Tennessee Valley, USA Partnership for Training in Environmental Education and Management," *International Research in Geographical and Environmental Education* 3, no. 2 (1994): 77–80.

McLeod, Mary. "The Country Is the Other City of Tomorrow?: Le Corbusier's Ferme Radieuse and Village Radieux." In *Food and the City: Histories of Culture and Cultivation*, edited by D. Imbert, 99–122. Washington, DC: Dumbarton Oaks Research Library and Collection, 2015.

McAlister, Melani. "'The Common Heritage of Mankind': Race, Nation, and Masculinity in the King Tut Exhibit." *Representations* 54 (1996): 80–103.

Nagata, Yūzō. *Tarihte âyânlar* [History of the Ayans]. Vol. 176. Turk Tarih Kurumu Basmevi, Istanbul 1997.

Naleszkiewicz, Wladimir. "Technical Assistance of the American Enterprises to the Growth of the Soviet Union, 1929–1933." *The Russian Review* 25, no. 1 (1966): 54–76.

Neyzi, Leyla. "Remembering Smyrna/Izmir: Shared History, Shared Trauma." *History and Memory* 20, no. 2 (2008): 106–127.

Norton, Charles E. "The Work of the Archaeological Institute of America: An Address." *American Journal of Archaeology* 4, no. 1 (1900): 1–16.

Oltermann, Philip, and Constanze Letsch, "Turkey Recalls Ambassador after German MPs' Armenian Genocide Vote," *The Guardian*, June 2, 2016.

Ousterhout, Robert, and Renata Holod, eds. *Osman Hamdi Bey ve Amerikalılar Arkeoloji, Diplomasi, Sanat: Osman Hamdi Bey & The Americans Archaeology, Diplomacy, Art*. Pera Müzesi Yayınları. Sergi Kataloğu / Exhibition Catalogue 51. İstanbul: Suna ve İnan Kıraç Vakfı, Pera Müzesi, 2011.

Özağaç, S. "Cumhuriyet Dönemi Türk Haritacılık Tarihi [The History of Turkish Cartography in the Republican Era]." *Yüksek Lisans Tezi, Ankara: TC Ankara Üniversitesi Türk İnkılâp Tarihi Enstitüsü*, 2006.

Özcan, Feridun Cemil. "US Aid and Turkish Macroeconomic Policy: A Narration of the Aid Bargain Process in the 1946–1958 Period." *The Turkish Yearbook of International Relations* 34 (2003): 119–136.

Özçatalbaş, Orhan. "Current Status of Advisory and Extension Services for Organic Agriculture in Europe and Turkey." In *Organic Agriculture towards Sustainability*, edited by V. Pilipavicius, 1301–1075. London: InTechOpen, 2014.

Özdoğan, Mehmet. *Lower Euphrates Basin 1977 Survey*. Middle East Technical University Keban and Lower Euphrates Projects Director, 1977.

Özen, Hayriye, and Şükrü Özen. "Interactions in and between Strategic Action Fields: A Comparative Analysis of Two Environmental Conflicts in Gold-Mining Fields in Turkey." *Organization & Environment* 24, no. 4 (2011): 343–363.

Özen, Hayriye, and S. Özen. "Public Policies and Social Movements: The Influences of Protest Movements on Mining Policy in Turkey." *Review of Public Administration* 43, no. 2 (2010): 33–64.

Özen, Şükrü, and Hayriye Özen. "Peasants against MNCs and the State: The Role of the Bergama Struggle in the Institutional Construction of the Gold-Mining Field in Turkey." *Organization* 16, no. 4 (2009): 547–573.

Özgen, İlknur, Jean Öztürk, and Machteld Johanna Mellink. *The Lydian Treasure: Heritage Recovered*. Published by Uğur Okman for Republic of Turkey, Ministry of Culture, General Directorate of Monuments and Museums, 1996.

Özyürek, Esra. "Miniaturizing Atatürk Privatization of State Imagery and Ideology in Turkey." *American Ethnologist* 31, no. 3 (2004): 374–391.

Padgett, J. Michael. "The Collections of Ancient Art: The Early Years." *Record of the Art Museum, Princeton University* 55, no. 1/2 (1996): 107–124.

Palancıoğlu, H. Mustafa, Abdurrahman Geymen, and Mehmet Alkan. "The Turkish Cadastral Information System and Lessons Learned." *Scientific Research and Essays* 5, no. 7 (2010): 625–633.

Parmar, Inderjeet. *Foundations of the American Century: The Ford, Carnegie, and Rockefeller Foundations in the Rise of American Power*. London: Columbia University Press, 2012.

Patrick, Andrew. *America's Forgotten Middle East Initiative: The King–Crane Commission of 1919*. Vol. 80. IB Tauris, 2015.

Patterson, Thomas C. "The Last Sixty Years: Toward a Social History of Americanist Archeology in the United States." *American Anthropologist* 88, no. 1 (1986): 7–26.

Pauly, Louis W. "Good Governance and Bad Policy: The Perils of International Organizational Overextension." *Review of International Political Economy* 6, no. 4 (1999): 401–424.

Pauly, Louis W. *Who Elected the Bankers? Surveillance and Control in the World Economy*. Ithaca, NY: Cornell University Press, 1997.

Pauly, Louis W. "The League of Nations and the Foreshadowing of the International Monetary Fund." *Essays in International Finance*, no. 201, Princeton University, International Finance Section, (1996): 1–47.

Pemberton, Jo-Anne. "The Changing Shape of Intellectual Cooperation: From the League of Nations to UNESCO." *Australian Journal of Politics & History* 58, no. 1 (2012): 34–50.

Pedersen, Susan. "Back to the League of Nations." *American Historical Review* 112, no. 4 (2007): 1091–1117.

Pedersen, Susan. *The Guardians: The League of Nations and the Crisis of Empire*. New York: Oxford University Press, 2015.

Peterson, Frederick J. "The Facade of Humanitarian Intervention for Human Rights in a Community of Sovereign Nations." *Arizona Journal of International and Comparative Law* 15 (1998): 871–904.

Phillips, Sarah T. *This Land, This Nation: Conservation, Rural America, and the New Deal*. New York: Cambridge University Press, 2007.

Plets, Gertjan. "Heritage Bureaucracies and the Modern Nation State. Towards an Ethnography of Archaeological Systems of Government." *Archaeological Dialogues* 23, no. 2 (2016): 193–213.

Plets, Gertjan. "Heritage Statecraft: When Archaeological Heritage Meets Neoliberalism in Gazprom's Resource Colonies, Russia." *Journal of Field Archaeology* 41, no. 3 (2016): 368–383.

Pollock, Susan. "Archaeology and Contemporary Warfare." *Annual Review of Anthropology* 45 (2016): 215–231.

Pribilsky, J. (2009). Development and the "Indian Problem" in the Cold War Andes: Indigenismo, Science, and Modernization in the Making of the Cornell-Peru Project at Vicos. *Diplomatic History* 33(3): 405–426.

Rajak, Dinah. *In Good Company: An Anatomy of Corporate Social Responsibility*. Stanford, CA: Stanford University Press, 2011.

Raleigh, Donald J. *Provincial Landscapes: Local Dimensions of Soviet Power, 1917–1953*. Pittsburg: University of Pittsburgh Press, 2001.

Ramage, Andrew, Paul T. Craddock, and Michael R. Cowell. *King Croesus' Gold: Excavations at Sardis and the History of Gold Refining*. Archaeological Exploration of Sardis Monograph 11. London: British Museum Publications, 2000.

Rassweiler, Anne Dickason. *The Generation of Power: The History of Dneprostroi*. Oxford University Press on Demand, 1988.

Ratté, Christopher. "Not the Tomb of Gyges." *Journal of Hellenic Studies* 114 (1994): 157–161.

Ratté, Christopher. *Lydian Architecture: Ashlar Masonry Structures at Sardis*. Archaeological Exploration of Sardis Monograph 5. Cambridge, MA: Harvard University Press, 2011.

Reitz, Louis P. "New Wheats and Social Progress." *Science* 169, no. 3949 (1970): 952–955.

Ricoeur, Paul. "Memory and Forgetting." In *Questioning Ethics: Contemporary Debates in Philosophy*, edited by M. Dooley and R. Kearney, 5–11. New York: Routledge, 1999.

Rizzo, Ilde, and Anna Mignosa, eds. *Handbook on the Economics of Cultural Heritage*. Edward Elgar Publishing, 2013.

Robbins, Peter T., and Ben Crow. "Engineering and Development: Interrogating Concepts and Practices." *Journal of International Development* 19, no. 1 (2007): 75–82.
Roberts, Timothy Mason. "Commercial Philanthropy: ABCfm Missionaries and the American Opium Trade." *The Role of the American Board in the World: Bicentennial Reflections on the Organization's Missionary Work, 1810–2010* (2012): 27.
Robinson, Richard D. "Tractors in the Village—A Study in Turkey." *Journal of Farm Economics* 34, no. 4 (1952): 451–462.
Rosenstein–Rodan, Paul Narziss. "International Aid for Underdeveloped Countries." *Review of Economics and Statistics* 43, no. 2 (1961): 107–138.
Rossow, Robert. "The Professionalization of the New Diplomacy." *World Politics* 14, no. 04 (1962): 561–575.
Roosevelt, Christopher H. *The Archaeology of Lydia, from Gyges to Alexander.* Cambridge, UK: Cambridge University Press, 2009.
Roosevelt, Christopher H., and Christina Luke. "The Story of a Forgotten Kingdom? Survey Archaeology and the Historical Geography of Central Western Anatolia in the Second Millennium BC." *European Journal of Archaeology* 20, no. 1 (2017): 120–147.
Roosevelt, Christopher H., and Christina Luke. "Mysterious Shepherds and Hidden Treasures: The Culture of Looting in Lydia, Western Turkey." *Journal of Field Archaeology* 31, no. 2 (2006): 185–198.
Roosevelt, Christopher. H., and Christina Luke. "Central Lydia Archaeological Survey: 2006 Results." *Araştırma Sonuçları Toplantısı* 25, no. 3 (2008): 305–326.
Roosevelt, Christopher. H., and Christina Luke. "Central Lydia Archaeological Survey: 2007 Results." *Araştırma Sonuçları Toplantısı* 26, no. 2 (2009): 433–450.
Roosevelt, Christopher. H., and Christina Luke. "Central Lydia Archaeological Survey: 2008 Results." Araştırma Sonuçları Toplantısı 27, no. 2 (2010): 1–24.
Roosevelt, Christopher. H., and Christina Luke. "Central Lydia Archaeological Survey: 2009 Results." *Araştırma Sonuçları Toplantısı* 28, no. 3 (2011): 55–74.
Roosevelt, Christopher. H., and Christina Luke. "Central Lydia Archaeological Survey: 2010 Results." *Araştırma Sonuçları Toplantısı* 29, no. 1 (2012): 383–400.
Roosevelt, Christopher. H., and Christina Luke. "The Central Lydia Archaeological Survey: 2011 Work at Kaymakçı and in the Marmara Lake Basin." *Araştırma Sonuçları Toplantısı* 30, no. 1 (2013): 237–254.
Roosevelt, Christopher H., Christina Luke, Peter Cobb, Caitlin O'Grady, and Bradely Sekedat. "The Central Lydia Archaeological Survey: 2012 Work at Kaymakçı and in the Marmara Lake Basin." *Araştırma Sonuçları Toplantısı* 31, no. 1 (2014): 333–355.
Roosevelt, Christopher H., Christina Luke, Bradely Sekedat, and Peter Cobb. "The Central Lydia Archaeological Survey: 2013 Work at Kaymakçı and in the Marmara Lake Basin." *Araştırma Sonuçları Toplantısı* 32, no. 2 (2015): 239–258.
Roosevelt, Christopher H., Christina Luke, and Sinan Ünlüsoy. "Kaymakçı Arkeoloji Projesi: 2014 Yılı Kazı Sonuçları." *Kazı Sonuçları Toplantısı* 37, no. 2 (2016): 243–268.
Roosevelt, Christopher H., Sinan Ünlüsoy, and Christina Luke. "Kaymakçı Arkeoloji Projesi: 2015 Yılı Kazı Sonuçları." *Kazı Sonuçları Toplantısı* 38, no. 2: 563–586.
Roy, Ananya. *Poverty Capital: Microfinance and the Making of Development.* London: Taylor and Francis, 2010.

Rydell, Robert W. *All the World's a Fair: Visions of Empire at American International Expositions, 1876–1916*. Chicago: University of Chicago Press, 2013.

Sarc, Ömer Celâl. "Economic Policy of the New Turkey." *Middle East Journal* 2, no. 4 (1948): 430–446.

Savino, Melania. "Connecting Sites and Images: Archeology as Controversial Knowledge in Modern Izmir." *History of Science* 55, no. 3 (2017): 364–382.

Saydam, Coşkun, Mustafa Öğüt, and Mustafa Copcu. "Ege Bölgesinde Yetiştirlilen Meksika Kaynaklı Buğdayların Hastaklıklarla Ilgisi ve kuru tohum ilaçlamasının sürme gücüne etkisi üzerinde çalışmalar [Studies on Diseases of Mexican Wheat Grown in the Aegean Region and the Effect of Pesticides on Dry Seed Germination]." *Bitki Koruma Bülteni Cilt* 14, no. 3 (1974): 151–180.

Scott, James C. *Seeing Like a State: How Certain Schemes to Improve the Human Condition Have Failed*. New Haven, CT: Yale University Press, 1998.

Schapsmeier, Edward L., and Frederick H. Schapsmeier. "Farm Policy from FDR to Eisenhower: Southern Democrats and the Politics of Agriculture." *Agricultural History* 53, no. 1 (1979): 352–371.

Schofield, Richard. "International Boundaries and Borderlands in the Middle East: Balancing Context, Exceptionalism and Representation." *Geopolitics* (2017): 1–24.

Sekedat, Bradley M. "X-Ray Fluorescence and Stable Isotope Analysis of Marble in Central Lydia, Western Turkey." *Oxford Journal of Archaeology* 35, no. 4 (2016): 369–388.

Selznick, Philip. *TVA and the Grass Roots: A Study in the Sociology of Formal Organization*. Vol. 3. Berkeley, CA: University of California Press, 1949.

Seydi, Süleyman. "Turkish-American Relations and the Cuban Missile Crisis, 1957–63." *Middle Eastern Studies* 46, no. 3 (2010): 433–455.

Shade, Lindsay. "Sustainable Development or Sacrifice Zone? Politics below the Surface in Post-Neoliberal Ecuador." *The Extractive Industries and Society* 2, no. 4 (2015): 775–784.

Shaw, Claire. "A Fairground for 'Building the New Man:' Gorky Park as a Site of Soviet Acculturation." *Urban History* 38, no. 02 (2011): 324–344.

Shaw, Wendy M. K. *Possessors and Possessed: Museums, Archaeology, and the Visualization of History in the Late Ottoman Empire*. Los Angeles: University of California Press, 2003.

Shokr, Ahmad. "Hydropolitics, Economy, and the Aswan High Dam in Mid-Century Egypt." *The Arab Studies Journal* 17, no. 1 (2009): 9–31.

Smith, Claire, Heather Burke, Cherrie de Leiuen, and Gary Jackson. "The Islamic State's Symbolic War: Da'esh's Socially Mediated Terrorism as a Threat to Cultural Heritage." *Journal of Social Archaeology* 16, no. 2 (2016): 164–188.

Sneddon, Chris *Concrete Revolution: Large Dams, Cold War Geopolitics, and the US Bureau of Reclamation*. Chicago: University of Chicago Press, 2015.

Sneddon, Chris, and Coleen Fox. "The Cold War, the US Bureau of Reclamation, and the Technopolitics of River Basin Development, 1950–1970." *Political Geography* 30, no. 8 (2011): 450–460.

Sora, Steven. *Secret Societies of America's Elite: From the Knights Templar to Skull and Bones*. Rochester, VT: Inner Traditions/Bear & Co., 2003.

Staples, Amy L. S. "Seeing Diplomacy through Banker's Eyes: The World Bank, the Anglo-Iranian Oil Crisis, and the Aswan High Dam." *Diplomatic History* 26, no. 3 (2002): 397–418.

Starr, Fiona. *Corporate Responsibility for Cultural Heritage: Conservation, Sustainable Development, and Corporate Reputation*. Vol. 4. New York: Routledge, 2013.

Steffen, Will, Jacques Grinevald, Paul Crutzen, and John McNeill. "The Anthropocene: Conceptual and Historical Perspectives." *Philosophical Transactions of the Royal Society of London A: Mathematical, Physical and Engineering Sciences* 369, no. 1938 (2011): 842–867.

Steffensen-Bruce, Ingrid A. *Marble Palaces, Temples of Art: Art Museums, Architecture, and American Culture, 1890–1930*. Lewisburg: Bucknell University Press, 1998.

Stein, Sarah Abrevaya. *Extraterritorial Dreams: European Citizenship, Sephardi Jews, and the Ottoman Twentieth Century*. Chicago: University of Chicago Press, 2016.

Stinespring, W. F. "Jerash in the Spring of 1934." *Bulletin of the American Schools of Oriental Research* 57 (1935): 3–9.

Stokstad, Erik. "Turkish Government Shuts Down Important Archaeological Dig, Apparently to Punish Austria," *Science*, September 6, 2016.

Stuart, James, Nicholas Revett, and Frank Salmon. *The Antiquities of Athens: Measured and Delineated by James Stuart, FRS and FSA, and Nicholas Revett, Painters and Architects*. New York: Princeton Architectural Press, 2008 [1762].

Tansuğ, Feryal. "The Greek Community of Izmir/Smyrna in an Age of Transition: The Relationship between Ottoman Centre-Local Governance and the Izmir/Smyrna Greeks, 1840–1866." *British Journal of Middle Eastern Studies* 38, no. 01 (2011): 41–72.

Takim, Abdullah, and Ensar Yilmaz. "Economic Policy during Atatürk's Era in Turkey (1923–1938)." *African Journal of Business Management* 4, no. 4 (2010): 549.

Tanyeri-Erdemir, Tuğba. "Archaeology as a Source of National Pride in the Early Years of the Turkish Republic." *Journal of Field Archaeology* 31, no. 4 (2006): 381–393.

Thompson, Dorothy. "The Great Productive Possibilities of Turkey." *Daily Boston Globe*, June 1, 1949.

Tillman, Seth P. *Anglo-American Relations at the Paris Peace Conference of 1919*. Princeton: Princeton University Press, 2015 [1961], 365–368.

Togan, Sübidey. "Turkey: Toward EU Accession." *World Economy* 27, no. 7 (2004): 1013–1045.

Toğrol, Beglan B. "The Kuban Cossacks of Manyas Turkey and 'The Cossacks' of Leo Tolstoy." *Psikoloji Çalışmaları Dergisi* (1987): 125–146.

Tomas-Simin, Mirela, and Danica Glavas-Trbic. "Historical Development of Organic Production." *Ekonomika Poljoprivrede* 63, no. 3 (2016): 1083.

Toynbee, Arnold. *The Western Question in Greece and Turkey*. London: Constable and Company, 1922.

Tsakiridou, Corinna. "Ideology, Prejudice, and Ethnic Conflict in George Horton's 'The Blight of Asia': A Radical Narrative Reconsidered." *Turkish Studies Association Bulletin* 25, no. 1 (2001): 21–38.

Truffer, Bernhard, and Lars Coenen. "Environmental Innovation and Sustainability Transitions in Regional Studies." *Regional Studies* 46, no. 1 (2012): 1–21.

Tusan, Michelle. *Smyrna's Ashes: Humanitarianism, Genocide, and the Birth of the Middle East*. Los Angeles: University of California Press, 2012.

Türkdoğan, Orhan. "Türkiye'de Göçler, Sorunları ve Çözüm Yolları [Migrations in Turkey: Problems and Solutions]," *İktisat Fakültesi Mecmuası* 35, no. 1–4 (1975): 40–67.

UN Global Compact. Accessed March 10, 2017. https://www.unglobalcompact.org/what-is-gc.

Unsal, Burcu, and Hasan Yazicigil. "Assessment of Open Pit Dewatering Requirements and Pit Lake Formation for the Kışladağ Gold Mine, Uşak, Turkey." *Mine Water and the Environment* 35, no. 2 (2016): 180–198.

Uzakgören, Yasemin. "Design Criteria for Increasing Environmental Performance of Recreational Areas: Analysis of Kordon/Izmir." Master's thesis, Izmir Institute of Technology, 2010.

VanderLippe, John M. *The Politics of Turkish Democracy: Ismet Inonu and the Formation of the Multi-Party System, 1938–1950*. Albany: SUNY Press, 2012.

Veeser, Cyrus. "A Forgotten Instrument of Global Capitalism? International Concessions, 1870–1930." *International History Review* 35, no. 5 (2013): 1136–1155.

Vrdoljak, Ana Filipa. "3 Self-Determination and Cultural Rights." In *Cultural Human Rights*, edited by F. Francioni and M. Scheinin, 41–78. Leiden: Brill, 2008.World Heritage and Extractive Industries, available from http://whc.unesco.org/en/extractive-industries/, accessed on April 25, 2017.

Watenpaugh, Heghnar Zeitlian. "Preserving the Medieval City of Ani: Cultural Heritage between Contest and Reconciliation." *Journal of the Society of Architectural Historians* 73, no. 4 (2014): 528–555.

Ward, Geoffrey C. *Before the Trumpet: Young Franklin Roosevelt, 1882–1905*. New York: Vintage, 2014.

Ward, Stephen V. "Soviet Communism and the British Planning Movement: Rational Learning or Utopian Imagining?" *Planning Perspectives* 27, no. 4 (2012): 499–524.

Weinstein, James. *The Corporate Ideal in the Liberal State: 1900–1918*. Boston: Beacon Press, 1968.

Wilber, Donald Newton. *Adventures in the Middle East: Excursions and Incursions*. Princeton, NJ: Darwin Press, 1986.

Williams, E. T. "The Conflict between Autocracy and Democracy." *The American Journal of International Law* 32, no. 4 (1938): 663–679.

Wilson, Christopher S. *Beyond Anitkabir: The Funerary Architecture of Atatürk: The Construction and Maintenance of National Memory*. London: Routledge, 2016.

Winder, Gordon M. *The American Reaper: Harvesting Networks and Technology, 1830–1910*. London: Routledge, 2016.

Winder, Gordon M. "A Trans-National Machine on the World Stage: Representing McCormick's Reaper through World's Fairs, 1851–1902." *Journal of Historical Geography* 33, no. 2 (2007): 352–376.

Wood, David, ed. *On Paul Ricoeur: Narrative and Interpretation*. New York: Routledge, 2002.

Wood, Jason, and B. Nilgün Öz. "The Turkish Amusement Park: Modernity, Identity and Cultural Change in the Early Republic." In *The Amusement Park: History, Culture and the Heritage of Pleasure*, edited by J. Wood, 98–119. London: Routledge.

Wood, Robert. "From the Marshall Plan to the Third World." In *The Origins of the Cold War: An International History*, 2nd ed., edited by M. P. Leffler and D. S. Painter, 239–250. New York: Routledge, 2005.

World Bank. "Using Data to Tell the Artisanal and Small-Scale Mining Story," January 30, 2017 (accessed March 2017). http://www.worldbank.org/en/news/feature/2017/01/30/using-data-to-tell-the-artisanal-and-small-scale-mining-story.

World of Organic Agriculture 2017 report by IFOAM EU Group: European Union Group of IFOAM—Organics International. Accessed February 20, 2017. http://www.organic-world.net/yearbook/yearbook-2017/pdf.html.

Wunsch, Natasha. "Beyond Instrumentalisation: NGO Monitoring Coalitions in Croatia, Montenegro, and Serbia." *East European Politics* 31, no. 4 (2015): 452–467.

Wyatt, S. C. "Turkey: The Economic Situation and the Five-Years Plan." *International Affairs (Royal Institute of International Affairs 1931–1939)* 13, no. 6 (1934): 826–844.

Yayla, Hilmi Erdoğan. "Operating Regimes of the Government: Accounting and Accountability Changes in the Sultan Süleyman Waqf of the Ottoman Empire (The 1826 Experience)." *Accounting History* 16, no. 1 (2011): 5–34.

Yegül, F. "Yağmalanan Sardes ve Türk Arkeolojisinin Geçmişle ilk Hesaplaşması, 1922–1925 [Plundered Sardis and a First Accounting of Turkish Archaeology's Past, 1922–1925]." In *Orhan Bingol 67. Yas Armagani*, edited by G. Kokdemir, 711–726. Ankara: Ankara University, 2013.

Yegül, Fikret. "From the Lofty Halls of Academia to the Dusty Hills of Anatolia: Howard Crosby Butler and the First Sardis Expedition through Peace and War, 1909–1926." In *Perceptions of the Past in the Turkish Republic: Classical and Byzantine Periods*, edited by S. Redford and N. Ergin, Ancient Near Eastern Studies, Supplement 31, 57–100. Leuven: Peeters, 2010.

Yegül, Fikert. "Architect Throwing Stone over Column 7 of Temple of Artemis." In *The City of Sardis: Approaches in Graphic Recording*, edited by C. H. Greenewalt, 74–75. Cambridge, MA: Harvard University Art Museum, 2003.

Yegül, Fikret. K. *Gentlemen of Instinct and Breeding: Architecture at the American Academy in Rome, 1894–1940*. New York: Oxford University Press, 1991.

Yegül, Fikret K., with contributions by Mehmet C. Bolgil, and Clive Foss. *The Bath-Gymnasium Complex at Sardis*. Vol. 3. Cambridge, MA: Harvard University Press, 1986.

Yiğit-Turan, Burcu. "Modernist Landscapes of Ankara." *Journal of Landscape Architecture* 11, no. 2 (2016): 14–25.

Yılmaz, Arı. "Manyas Gölü'nün kültürel ekolojisi: Tarihi süreçte adaptasyon ve değişimi [The Cultural Ecology of Lake Manyas: Adaptation and Change in Historical Perspective]." *Türk Coğrafya Dergisi* 40 (2003): 75–97.

Zacharia, Katerina. "Postcards from Metaxas' Greece: The uses of classical antiquity in tourism photography." In *Re-imagining the Past: Antiquity and Modern Greek Culture,* edited by D. Tziovas, 186–208. New York: Oxford University Press.

Zajicek, Taylor Craig. "Modern Friendship: The 'New Turkey' and Soviet Cultural Diplomacy, 1933–1934." PhD diss., University of Washington, 2014.

Zandi-Sayek, Sibel. *Ottoman Izmir: The Rise of a Cosmopolitan Port, 1840–1880*. Minneapolis, MN: University of Minnesota Press for Architecture, 2012.

INDEX